Chances Are ...

Chances Are ...

Making Probability and Statistics Fun to Learn and Easy to Teach

Nancy Pfenning, Ph.D.
Carnegie Mellon University Institute for Talented Elementary Students (C–MITES)
and University of Pittsburgh

Copyright ©1998, Nancy Pfenning

ISBN 1-882664-35-3

Cover design by Libby Lindsey

PRUFROCK
PRESS INC.™

P.O. Box 8813, Waco, TX 76714
(254) 756-3337 ▼ fax (254) 756-3339
http://www.prufrock.com

For my parents

Acknowledgments

Thanks to Ann Lupkowski-Shoplik for accommodating me as teacher in her C-MITES summer program for gifted students, for which I developed this book. Also to Susan Barie, who welcomed me into her wonderful sixth grade math class at Frick International Studies Academy, enabling me to give various hands-on activities a trial run. Having taught probability and statistics for quite a few years at the University of Pittsburgh and incubated ideas all along, I often felt this book was practically "writing itself." It did not, however, prepare itself for submission; thanks to my husband Frank, who helped so much to polish the readability and look of the original manuscript, making it easier for me to polish the contents. I am grateful to him and our three children for their encouragement throughout, and their willingness to serve as guinea pigs on occasion. Finally, I thank my parents, Bill and Elena Gormley, for whom the urge to improve one's mind with learning is just as strong as the need to pass on knowledge with teaching. This, and so much more, they have passed on to me.

Contents

Preface

Probability—the study of random behavior—and statistics—the formal treatment of data—began in the 18th century with people's interest in the results of games of chance, such as cards, dominoes, coin flips, and so forth. As sciences, they lend themselves to hand-in-hand study. Mathematicians tend to consider the concepts of probability to be interesting in their own right, but students may not find the study gratifying unless they can see probability's usefulness in statistical applications. On the other hand, even the most basic concepts of statistics would be difficult to present in a coherent way if some groundwork were not laid by establishing the rules of probability.

The beauty of probability is that such a rich theory emerges from just a few basic assumptions. The intricacy of this theory, which is so pleasing to mathematicians, can be overwhelming to students. The purpose of the "Probability" section of this book is to present an overview with the minimal amount of theory necessary to understand the "Statistics" section that follows and to make it as intuitive as possible by working through straightforward examples and by including simple, fun experiments.

The limitless expanse of numbers and contexts which can be tackled with statistical methods accounts for statistics' undisputed usefulness in all sorts of other disciplines, from medicine to sports. The current fashion is to work immediately with real-life data sets from these disciplines, rather than with those tailored to fit the purpose of teaching. The drawback to this approach is the fact that most real-life statistical problems are quite complex and most real-life numbers are messy. Simple statistical concepts may be lost on students who are struggling with the nuances of a word problem or with long strings of digits to be punched into a calculator or a computer. The purpose of the "Statistics" section of this book is to examine the most basic features of statistical theory in very uncomplicated settings with very manageable numbers. Again, experiments are included so that students can see how concrete results tie in to the abstract theory.

Probability and statistics may be the horror of many college students, but if they are trimmed to include only the essentials and explained with a minimum of exotic symbols, they are easily within the grasp of interested middle school or even elementary school students. This book can serve as an introduction for any beginner, from gifted students who would like to broaden their horizons while keeping in step with their school's conventional math curriculum, to nervous college students who are facing their first "prob-and-stats" requirement. High school students may use it as enrichment or as preparation for Advanced Placement Statistics or future college courses. It is hoped that through such an introduction, the reader will be able to share in the excitement that first inspired scholars to explore chance behavior.

Overview

In the Introduction, probability and statistics as sciences are introduced, along with their connection to one another and the approach we intend to use studying them.

Counting lays a foundation for the study of probability by presenting ways to count up all the possibilities in various situations. Once this has been accomplished, we see in Probabilities how to solve for probabilities, taking into account all the possibilities in a situation. We derive the most basic rules for manipulating probabilities, including special and general addition and multiplication rules. In Probability Distributions, we proceed from solving for individual probabilities to solving for probability distributions, which show us the long-run patterns occurring in random behavior. Attention is given to the difference between independent and dependent events by examining distributions for sampling with and without replacement. Statistical Inference, which may be omitted for students at lower levels, introduces the process of hypothesis testing for discrete probability distributions such as we have seen in Probability Distributions.

The statistics half of the book begins in Beginning Statistics, where numerical and graphical summaries of data are presented, including, of course, the mean of a data set. In Sample Mean and Sample Proportion the notion of a statistic as something measured from a sample is introduced. We examine two statistics—sample mean and sample proportion—and relate them to the unknown mean or proportion of the whole population from which the sample was taken. In Histograms and the Normal Distribution we relate the patterns for particular measurements of a variable to the patterns for all possible values of the variable. This leads us to the most important pattern of behavior of a variable—the normal distribution. In order to summarize a distribution, besides using the mean to tell where it is centered, we use standard deviation to tell about its spread. Standard Deviation presents not only the calculation of standard deviation, but also its physical interpretation, its role in the normal distribution, and, finally, the way sample size affects the standard deviation of the distributions of sample mean and sample proportion. Handling Data in Different Forms begins with a review of numerical and graphical summaries for various forms of data already presented and then explores the summary of relationships between two numerical variables. Statistical Inference II, like Statistical Inference I, is recommended for students who are ready to study the processes of statistical inference—specifically, finding confidence intervals and performing hypothesis tests. The role of sample size in making estimates is discussed, along with the impact of the Central Limit Theorem.

Introduction

Playing a game with dice, you may ask yourself, "What are my chances of getting a seven on the next roll?" or playing cards, you may wonder, "How likely would it be for the next card I pick to be a diamond?"

Shifting to a totally different context, suppose a doctor is informed by his assistant that a new patient has a pulse rate of 113. Should he be alarmed? If a fan wants to estimate the overall proportion of hits a baseball player gets when he is at bat, does it make sense to refer to the proportion of hits he gets in a particular game? Would the estimate be better if it were based on his proportion of hits in several games instead of just one?

Analyzing the results of dice rolls and card selections, because such results are truly random, is an ideal way to approach the science of probability, which is the study of random behavior. Although each individual dice roll or card selection is purely a matter of chance, we still know there will be certain patterns emerging after many dice rolls or card selections. The science of probability teaches us about such patterns in an organized way.

The connection between probability and statistics is somewhat complicated. Studying the patterns that occur in purely random behavior (such as dice rolls and card selections) helps us to understand the patterns that occur in real-life data sets (such as lists of patients' pulse rates or baseball players' batting averages). In statistics, means or proportions from samples are often used as estimates. With the help of probability theory, statisticians can determine how accurate these estimates are for the mean or proportion of the larger groups from which the samples were taken. At the same time, they can discover what kinds of samples yield the best estimates.

Pioneers in probability and statistics who lived in the 18th century did more than just think and figure with pencil and paper. They actually used dice or cards or coin flips to explore first-hand the patterns that occur naturally in random behavior. Like these original students of probability and statistics, we will experiment as we go along and see that the various principles hold not just in theory but also in practice. Along the way, we will find that probability and statistics, which are among the most useful subjects in modern education, also happen to be fascinating. Have fun with them!

Counting

Beginning Experiment: Counting Color Arrangements

Materials: stickers, one box each of six different colors, e.g. from office supply store; or pens/crayons/markers of six different colors; 11" x 17" paper or pairs of 8½" x 11" sheets taped together

After finding all possible ways to place the colors according to each of the instructions below, count them up for each different task.

1. Use marking pens or stickers which are **three** different colors. In how many ways can you place **three** colors in a row? [Colors *may* be repeated in your arrangement.]
2. Use marking pens or stickers which are **four** different colors. In how many ways can you place **four** different colors in a row? [Colors may *not* be repeated.]
3. Use marking pens or stickers which are **five** different colors. In how many ways can you place any **two** different colors in a row? [Order *is* important: red then blue counts as different from blue then red.]
4. Use marking pens or stickers which are **six** different colors. In how many ways can you choose any **three** out of those six colors? [Order is *not* important: green then blue then black counts as the same as green then black then blue. Colors may *not* be repeated.]
5. (Optional) Use marking pens or stickers which are **two** different colors. In how many ways can you make an arrangement of **six** in a row, using **three** of one color and **three** of the other color?

Introduction

Often, determining a probability is simple as soon as all the possibilities in a given situation have been taken into account. Learning how to count up all the possible ways for an event to occur is a good way to begin studying the science of probability. Most situations of interest to us actually fall into one of just a few patterns. We will begin with one of the simplest types of situations.

Two-Stage Process

Question   For her birthday, Marina will invite her friend to a restaurant, and then go to a movie afterward. If there are three good restaurants to choose from and four movies that seem promising, in how many different ways can Marina celebrate her birthday?

Answer   If we label the restaurants R_1 through R_3, and the movies M_1 through M_4, we can list all the possibilities as follows:

$$
\begin{array}{ccc}
R_1M_1 & R_2M_1 & R_3M_1 \\
R_1M_2 & R_2M_2 & R_3M_2 \\
R_1M_3 & R_2M_3 & R_3M_3 \\
R_1M_4 & R_2M_4 & R_3M_4
\end{array}
$$

We see there are three possibilities for a restaurant, and for each of these possibilities there are four possibilities for a movie, totaling $3 \times 4 = 12$ possibilities. Instead of listing all the possibilities, we may use a **tree diagram** to depict them:

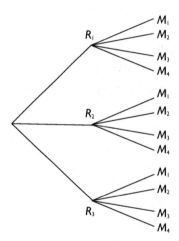

Counting up the total number of "leaves" ending at the right of the sideways tree is really the same as counting up all the possibilities in our list, starting with R_1M_1 and ending with R_3M_4.

[How many restaurants would be on your own list of favorites? How many movies are you interested in seeing? How many different possibilities would there be for you to go to one of these restaurants and then see one of these movies?]

Basic Rule   Suppose a process is made up of two stages where the first stage has n_1 possibilities, while the second stage has n_2 possibilities. Then the entire process has $n_1 \times n_2$ possibilities.

Example   Suppose a six-sided die is rolled twice. How many different possibilities can occur in this process?

Solution   There are six possibilities for the first roll of the die, and, for each of these, six possibilities for the second roll. Using our Basic Rule with $n_1 = 6$ and $n_2 = 6$, we find that altogether there are $6 \times 6 = 36$ possibilities.

Several-Stage Process

Question   For her birthday, Marina will invite a friend to one of three restaurants, and then go to one of four movies. After that, they will go for dessert at one of two ice cream shops. Now how many ways are there for Marina to spend her birthday?

Answer   There are 3×4 possible ways to go to a restaurant and then a movie. For each of these, there are 2 possible ways to go for dessert. Altogether there are $3 \times 4 \times 2 = 24$ possibilities, from $R_1M_1D_1$ to $R_3M_4D_2$. If we were to depict them in a tree diagram, there would be 24 "leaves" ending at the right.

Basic Rule	Suppose an entire process consists of several stages, with a given number of possibilities for each stage. Then we find the total number of possibilities for the entire process by multiplying the numbers of possibilities for all the stages together.

Example   Suppose a die is rolled three times. How many different possibilities can occur in this process?

Solution   Since each of the three rolls has six possible numbers, this is like a three-stage process where each stage has six possibilities. We apply our Basic Rule and find a total of $6 \times 6 \times 6 = 216$ possibilities altogether.

Permutations

A **permutation** is an arrangement of objects where the order is important.

Question   In how many different ways can we arrange the letters A, B, and C? In other words, how many permutations are there of the letters A, B, and C?

Answer   There are 3 possibilities for the first letter in our arrangement. For each of these, there are 2 possibilities for the second letter, because one letter has already been used. For each of these two-letter arrangements, there is only 1 remaining possibility for the third letter. Altogether, there are 3 x 2 x 1 = 6 possibilities. We could list the possibilities

ABC	BAC	CAB
ACB	BCA	CBA

or use a tree diagram and count six leaves ending on the right:

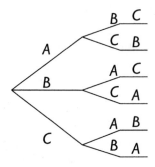

In fact, arranging the letters is like a three-stage process with 3 possibilities for the first, 2 for the second, and 1 for the third.

Example   How many different permutations are there of the letters A, B, C, D, and E?

Solution   This is like a five-stage process with 5 possibilities for the first, 4 for the second, 3 for the third, 2 for the fourth, and 1 for the fifth. Altogether there are 5 x 4 x 3 x 2 x 1 = 120 possibilities.

Notation for Factorials

Counting up all the possibilities in a probability problem often involves counting permutations which, in turn, involves multiplying several whole numbers in descending order. The **factorial** notation, written as an exclamation point, saves us from writing a long list of numbers and also helps us concentrate on the numbers which are most important in a problem. As an example, we will say "one factorial equals one" and write "1! = 1." Our general rule will be that 0! = 1 and n! = n x (n - 1) x (n - 2) x ... x 1 for any whole number n which is one or more. Here is a list of factorial values for the first few numbers:

4

0! = 1
1! = 1
2! = 2 x 1 = 2
3! = 3 x 2 x 1 = 6
4! = 4 x 3 x 2 x 1 = 24
5! = 5 x 4 x 3 x 2 x 1 = 120

and so on. Now we have an easy way to write a rule for counting permutations.

Basic Rule	There are *n*! ways to arrange *n* different objects. In other words, the number of permutations of *n* different objects is *n*!.

Example In how many ways can seven people stand in line to have their picture taken?

Solution There are 7! = 7 x 6 x 5 x 4 x 3 x 2 x 1 = 5,040 different permutations, or ordered arrangements, of those seven people.

Permutations of Objects Taken from a Larger Group

Question How many permutations are there consisting of five different letters taken from the entire alphabet?

Answer Since there are a total of 26 alphabet letters, there are 26 ways to pick the first letter in our ordered arrangement. For each of these, there are 25 ways to pick a different second letter. For each of these two-letter arrangements, there are 24 ways to pick a third

letter different from the first two. For each of these three-letter arrangements, there are 23 ways to pick a different fourth letter, and for each of these four-letter arrangements, there are 22 ways to pick a different fifth letter. Altogether there are

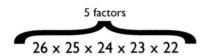

$$\overbrace{26 \times 25 \times 24 \times 23 \times 22}^{\text{5 factors}}$$

ways to pick five different alphabet letters.

Basic Rule	The number of permutations of *r* objects taken from *n* different objects is

$$\overbrace{n \times (n - 1) \times ... \times (n - r + 1)}^{\text{r factors}}$$

Example   Out of a group of seven people, in how many ways can we line up three of them to have their picture taken?

Solution   Using our basic rule with *r* = 3 and *n* = 7, we can say that the number of permutations of 3 people taken from seven different people is

$$\overbrace{7 \times 6 \times 5}^{\text{3 factors}} = 210$$

Combinations

A **combination** is a selection from a group of distinct objects where order is not important.

Question   How many *combinations* of three letters can we make from the letters A, B, C, D, and E?

Answer   First, we know there are $\overbrace{5 \times 4 \times 3}^{\text{3 factors}} = 60$ *permutations*, or *ordered arrangements* of three letters taken from five different letters:

Permutations

ABC	ABD	ABE	ACD	ACE	ADE	BCD	BCE	BDE	CDE
ACB	ADB	AEB	ADC	AEC	AED	BDC	BEC	BED	CED
BAC	BAD	BAE	CAD	CAE	DAE	CBD	CBE	DBE	DCE
BCA	BDA	BEA	CDA	CEA	DEA	CDB	CEB	DEB	DEC
CAB	DAB	EAB	DAC	EAC	EAD	DBC	EBC	EBD	ECD
CBA	DBA	EBA	DCA	ECA	EDA	DCB	ECB	EDB	EDC

But there are fewer combinations, or arrangements where order does not matter. In fact, the number of permutations of three letters is six times the number of combinations. This is because each combination of three letters is rearranged $3! = 6$ times when we count permutations. For example, the combination ABC, which takes up the first column above, includes the six permutations ABC, ACB, BAC, BCA, CAB, and CBA. To count combinations, we just divide the number of permutations by $3!$, and find altogether

$$\underbrace{\frac{\overbrace{5 \times 4 \times 3}^{\text{3 factors}}}{3 \times 2 \times 1}}_{3!} = \frac{5 \times 4}{2 \times 1} = 10$$

combinations of three letters taken from five letters.

Combinations

ABC   ABD   ABE   ACD   ACE   ADE   BCD   BCE   BDE   CDE

[Looking at our list of permutations above, we could count the number of combinations by counting the number of columns: there are 10 of them.]

Notation for Combinations

Because counting combinations is essential in the study of probability, we use the following special notation: For any non-negative whole numbers n and r (with r no greater than n), we write

$$\binom{n}{r} = \underbrace{\frac{\overbrace{n \times (n-1) \times \ldots \times (n-r+1)}^{r \text{ factors}}}{r \times (r-1) \times \ldots \times 2 \times 1}}_{r!}$$

This expression is called "n choose r" because it counts the number of ways to *choose r objects from n different objects*.

Basic Rule   The number of combinations of *r* objects taken from a group of *n* different objects is

$$\binom{n}{r}$$

Note that there is only one way to choose zero objects from any number of objects. The number of ways to choose one object from a group of objects is the same as the number of objects. Finally, there is only one way to choose all of the objects from a group. Thus,

$$\binom{n}{0} = 1, \qquad \binom{n}{1} = n, \qquad \binom{n}{n} = 1$$

for any positive whole number *n*.

Example   How many combinations of three can be taken from a group of seven people?

Solution   According to our basic rule, there are

$$\binom{7}{3} = \frac{\overbrace{7 \times 6 \times 5}^{3 \text{ factors}}}{\underbrace{3 \times 2 \times 1}_{3!}} = 7 \times 5 = 35$$

possible combinations of three people taken from seven.

[Notice that we have canceled the 6 in the numerator with 3 x 2 x 1 in the denominator. In general, canceling can make solving for combinations much simpler.]

Example   How many rolls of 36-exposure film would be needed if I wanted to photograph all possible combinations of 4 students taken from 20?

Solution The number of combinations of 4 students taken from 20 is

$$\binom{20}{4} = \frac{\overbrace{20 \times 19 \times 18 \times 17}^{4 \text{ factors}}}{\underbrace{4 \times 3 \times 2 \times 1}_{4!}} = 4{,}845$$

Dividing 4,845 by 36, we find that about 135 rolls would be needed.

Question In how many ways can we choose 2 boys and 1 girl from a group of 8 boys and 4 girls?

Answer There are

$$\binom{8}{2} = \frac{\overbrace{8 \times 7}^{2 \text{ factors}}}{\underbrace{2 \times 1}_{2!}} = 28$$

ways to choose 2 of the 8 boys. There are

$$\binom{4}{1} = 4$$

ways to choose 1 of the 4 girls.

Choosing 2 of the 8 boys and 1 of the 4 girls is a two-stage process in which the first stage can be done in 28 ways and the second stage in 4 ways. Altogether the process can be done in 28 × 4 = 112 ways.

In general, the number of ways to choose r_1 from n_1 objects and r_2 from n_2 objects is

$$\binom{n_1}{r_1} \times \binom{n_2}{r_2}$$

Example How many ways are there to choose 0 jelly doughnuts and 2 cream doughnuts from a bag containing 3 jelly doughnuts and 5 cream doughnuts?

Solution   There are

$$\binom{3}{0} \times \binom{5}{2} = 1 \times \underbrace{\overbrace{\frac{5 \times 4}{2 \times 1}}^{2 \text{ factors}}}_{2!} = 10$$

ways to choose 0 jelly doughnuts from 3 and 2 cream doughnuts from 5.

More Examples

Example   If three dice are rolled, in how many ways can there be three numbers all different?

Solution   There are 6 possibilities for the first die; for each of these, there are 5 possible ways for the second die to be different from the first, and for each of these there are 4 possible ways for the third die to be different from the first two. Altogether, there are 6 x 5 x 4 = 120 possibilities.

Example   If three dice are rolled, in how many ways can there be a pair of numbers the same and the other number different?

Solution   There are

$$\binom{3}{2} = 3$$

possible combinations of two dice which produce the pair:

	First Die	Second Die	Third Die
1.	⚁	⚁	
2.	⚁		⚁
3.		⚁	⚁

The number on these two dice may be any number from one to six, so there are

$$6 \times \binom{3}{2}$$

possible ways to get any pair. For each of these, there are five ways for the remaining die to be different from the pair. Altogether there are

$$6 \times \binom{3}{2} \times 5 = 90 \text{ possibilities.}$$

Example If three dice are rolled, in how many ways can there be three-of-a-kind?

Solution When all three dice are the same, they can all be any number from one to six, so there are 6 possibilities altogether.

Example If five dice are rolled (as is done in the game Yahtzee™), how many ways are there to get a "full house," that is, three-of-a-kind and a pair?

Solution There are

$$\binom{5}{3} = \overbrace{\underbrace{\frac{5 \times 4 \times 3}{3 \times 2 \times 1}}_{3!}}^{3 \text{ factors}} = 10$$

possible combinations of three dice from the five which produce three-of-a-kind:

	First Die	Second Die	Third Die	Fourth Die	Fifth Die		First Die	Second Die	Third Die	Fourth Die	Fifth Die
1.	⚁	⚁	⚁			6.	⚁			⚁	⚁
2.	⚁	⚁		⚁		7.	⚁	⚁	⚁		
3.	⚁	⚁			⚁	8.	⚁	⚁			⚁
4.	⚁		⚁	⚁		9.	⚁			⚁	⚁
5.	⚁		⚁		⚁	10.	⚁		⚁	⚁	⚁

They can be any number from one to six, so there are

$$6 \times \binom{5}{3}$$

ways to get three-of-a-kind. For each of these, the two remaining dice which produce the pair can be any of 5 possible numbers different from the three-of-a-kind. Altogether the number of possibilities for a full house are

$$6 \times \binom{5}{3} \times 5 = 6 \times 10 \times 5 = 300.$$

Example   How many ways can 3 blocks be chosen from 10 blocks?

Solution   There are

$$\binom{10}{3} = \frac{10 \times 9 \times 8}{3 \times 2 \times 1} = 120$$

ways to choose 3 blocks from 10 blocks.

◆   ◆   ◆

Example   Suppose 3 blocks are chosen from 10, where 6 are blue and 4 are not. In how many ways can there be 2 blue blocks chosen and 1 that is not blue?

Solution   The total number of ways to get 2 from 6 blue blocks and 1 from 4 non-blue blocks is

$$\binom{6}{2} \times \binom{4}{1} = \frac{6 \times 5}{2 \times 1} \times 4 = 15 \times 4 = 60$$

◆   ◆   ◆

Example   A single octave of white keys on a keyboard contains eight notes. How many "melodies" of three notes in a row are possible if notes may be repeated?

Solution   There are 8 x 8 x 8 = 512 possible ways to play three of the eight notes in a row, allowing repetition.

◆   ◆   ◆

Example   From an octave containing eight notes, how many melodies of three notes are possible if notes may not be repeated?

Solution   There are 8 x 7 x 6 = 336 possible ways to play three out of eight notes in a row, not allowing repetitions.

Example   From an octave containing eight notes, how many "chords" of three simultaneous notes are possible?

Solution   There are

$$\binom{8}{3} = \frac{8 \times 7 \times 6}{3 \times 2 \times 1} = 56$$

possible combinations of three notes to be played simultaneously. [If you have access to a piano or other keyboard, try playing all of them!]

◆   ◆   ◆

Example   Ordinary "double-six" dominoes are made with 2 squares of dots: each square contains anywhere from zero to six dots, so there are 7 possibilities for a single square. There is a domino for each possible combination of 2 squares of dots, and doubles are also included, but no two dominoes are alike. Why are there 28 dominoes in an ordinary "double-six" game?

Solution   There are

$$\binom{7}{2} = \frac{7 \times 6}{2 \times 1} = 21$$

different combinations of 2 from 7 square possibilities, and 7 double dominoes. Altogether, there must be 21 + 7 = 28 dominoes in a "double-six" game.

◆   ◆   ◆

Example   How many dominoes are there in a "double-nine" game, which has combinations and doubles of dots numbering zero through nine?

Solution   Since there are 10 possibilities for each square in a "double-nine" game, there are a total of

$$\binom{10}{2} + 10 = \frac{10 \times 9}{2 \times 1} + 10 = 45 + 10 = 55$$

different dominoes.

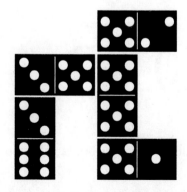

Special Combinations and Permutations (Optional)

Question How many combinations of 4 can be chosen from a group of 7 people?

Answer There are

$$\binom{7}{4} = \underbrace{\frac{\overbrace{7 \times 6 \times 5 \times 4}^{4 \text{ factors}}}{4 \times 3 \times 2 \times 1}}_{4!} = \underbrace{\frac{\overbrace{7 \times 6 \times 5}^{3 \text{ factors}}}{3 \times 2 \times 1}}_{3!} = 35$$

possible combinations of 4 people taken from 7.

Alternate Answer The number of combinations of 4 taken from 7 is actually the same as the number of combinations of 3 who are left *unchosen*, or

$$\binom{7}{3} = 35.$$

Basic Rule For n different objects and a whole number r between 0 and n,

$$\binom{n}{r} = \binom{n}{n-r}.$$

Example How many combinations are possible when 2 doughnuts are selected from 8? How many combinations are possible when 6 doughnuts are selected from 8?

Solution There are

$$\binom{8}{2} = \underbrace{\frac{\overbrace{8 \times 7}^{2 \text{ factors}}}{2 \times 1}}_{2!} = 28$$

possible combinations of 2 doughnuts selected from 8. There are

$$\binom{8}{6} = \binom{8}{8-6} = \binom{8}{2} = 28$$

possible combinations of 6 doughnuts selected from 8.

Question   How many permutations are there of the five letters A, B, C, C, C?

Answer   We know how to count the permutations if the three Cs were all different. To show that we are thinking about three different letters, we will call them C_1, C_2, and C_3, and count 5! = 120 permutations:

$ABC_1C_2C_3$	$AC_1BC_2C_3$	...	$C_1C_2C_3BA$
$ABC_1C_3C_2$	$AC_1BC_3C_2$	...	$C_1C_3C_2BA$
$ABC_2C_1C_3$	$AC_2BC_1C_3$	...	$C_2C_1C_3BA$
$ABC_2C_3C_1$	$AC_2BC_3C_1$	...	$C_2C_3C_1BA$
$ABC_3C_1C_2$	$AC_3BC_1C_2$	...	$C_3C_1C_2BA$
$ABC_3C_2C_1$	$AC_3BC_2C_1$	...	$C_3C_2C_1BA$

But since the three Cs are actually all alike, all six permutations in the first column should just be counted as one, namely, as ABCCC. All six permutations in the second column should be counted as ACBCC, and so on. Thinking of the three Cs as being different from one another causes us to count six times too many permutations, because 3! = 6 is the number of permutations of those three Cs. To get the correct answer, we divide 5! by 3!: There are $^5\!/_{3!}$ = 20 permutations of the five letters A, B, C, C, and C.

Question   How many permutations are there of the five letters A, A, C, C, and C?

Answer   From our answer above, we know there would be $^5\!/_{3!}$ = 20 permutations if the two As were different:

$$A_1A_2CCC \qquad A_1CA_2CC \qquad ... \qquad CCCA_2A_1$$

But since they are the same, we have to divide by the number of permutations of those two identical As—there are two of them, namely A_1A_2 and A_2A_1. Therefore, the number of permutations of the five letters A, A, C, C, and C is

$$^{5!}\!/_{(2! \times 3!)} = \binom{5}{2}.$$

Basic Rule	The number of permutations of *n* objects, where *r* of them are all the same as one another and the remaining *n - r* are also the same as one another, is $$\binom{n}{r}.$$

Example In how many ways can a sports team's season of six games result in four wins and two losses?

Solution The number of *permutations* of six games, where four are wins and two are losses, is $\frac{6!}{(4! \times 2!)} = 15$. Using *W* to denote a win and *L* to denote a loss, we can list all the possibilities as follows:

WWWWLL	WWLWWL	WLWWWL	WLLWWW	LWWLWW
WWWLWL	WWLWLW	WLWWLW	LWWWWL	LWLWWW
WWWLLW	WWLLWW	WLWLWW	LWWWLW	LLWWWW

Alternate Solution We can count how many *combinations* are possible when we know four of the six games have been won: the number of combinations of four games that are won taken from a group of six different games is

$$\binom{6}{4} = \frac{\overbrace{6 \times 5 \times 4 \times 3}^{4\ \text{factors}}}{\underbrace{4 \times 3 \times 2 \times 1}_{4!}} = 15$$

Listing combinations in which four games may be won again shows that there are 15 possibilities:

1st 2nd 3rd 4th	1st 2nd 4th 5th	1st 3rd 4th 5th	1st 4th 5th 6th	2nd 3rd 5th 6th
1st 2nd 3rd 5th	1st 2nd 4th 6th	1st 3rd 4th 6th	2nd 3rd 4th 5th	2nd 4th 5th 6th
1st 2nd 3rd 6th	1st 2nd 5th 6th	1st 3rd 5th 6th	2nd 3rd 4th 6th	3rd 4th 5th 6th

Finishing Experiment:
Counting Color Arrangements

Use what you have learned about several-stage processes, permutations, and combinations to count the number of possibilities for each of the tasks in the beginning experiment. If your answers are different now, make corrections as needed.

Exercise: Pascal's Triangle

1. The triangle shown below is called Pascal's triangle. Explain what patterns you see in the numbers, and complete at least one more row of the triangle.

$$
\begin{array}{ccccccccccc}
 & & & & & 1 & & & & & \\
 & & & & 1 & & 1 & & & & \\
 & & & 1 & & 2 & & 1 & & & \\
 & & 1 & & 3 & & 3 & & 1 & & \\
 & 1 & & 4 & & 6 & & 4 & & 1 & \\
1 & & 5 & & 10 & & 10 & & 5 & & 1
\end{array}
$$

2. Now note the pattern of the combinations in the triangle shown below, and solve for their values, keeping in mind that

$$
\binom{n}{r} = \frac{\overbrace{n \times (n-1) \times \dots \times (n-r+1)}^{r \text{ factors}}}{\underbrace{r \times (r-1) \times \dots \times 2 \times 1}_{r!}}
$$

$$
\begin{array}{ccccccccccc}
 & & & & & \binom{0}{0} & & & & & \\
 & & & & \binom{1}{0} & & \binom{1}{1} & & & & \\
 & & & \binom{2}{0} & & \binom{2}{1} & & \binom{2}{2} & & & \\
 & & \binom{3}{0} & & \binom{3}{1} & & \binom{3}{2} & & \binom{3}{3} & & \\
 & \binom{4}{0} & & \binom{4}{1} & & \binom{4}{2} & & \binom{4}{3} & & \binom{4}{4} & \\
\binom{5}{0} & & \binom{5}{1} & & \binom{5}{2} & & \binom{5}{3} & & \binom{5}{4} & & \binom{5}{5}
\end{array}
$$

You'll need to use the factorial values

0! = 1
1! = 1
2! = 2 x 1 = 2
3! = 3 x 2 x 1 = 6
4! = 4 x 3 x 2 x 1 = 24
5! = 5 x 4 x 3 x 2 x 1 = 120

Also, remember that

$$\binom{n}{0} = 1 \qquad \binom{n}{1} = n \qquad \binom{n}{n} = 1$$

for any whole number n greater than or equal to zero.

As an example, here are the values for the fourth row:

$$\binom{3}{0} = 1$$

$$\binom{3}{1} = 3$$

$$\binom{3}{2} = \frac{\overbrace{3 \times 2}^{\text{2 factors}}}{\underbrace{2 \times 1}_{2!}} = 3$$

$$\binom{3}{3} = 1$$

3. (Optional) Make up (and solve) a problem involving permutations. Make up (and solve) a problem involving combinations. Now that you have learned about permutations, combinations, and several-stage processes, can you think of a "better" name for a "combination lock"?

Probabilities

Introduction

The probability of an event tells us what chance it has of occurring. If an event could never occur, we agree that it has a probability of zero, and if it must occur, it has a probability of one. If a balanced coin is tossed, we say the probability of getting a head is 50% or .50 or ½ because it is just as likely as not for a head to appear. If we know that in the long run, an event will occur 20 times out of 100, then we say it has a probability of 20% or .20 or ⅕. Whether we use percentages or decimals or fractions to represent a probability usually depends on the problem setting.

Our everyday lives are full of situations that can be studied in the context of the science of probability. When we take a scientific approach to a situation, we may call it an *experiment*. The original probability experiments included dice rolls and card games, but we will examine many others, including such ordinary "experiments" as picking a doughnut from a bag or choosing a class member at random.

Basic Rules of Probability

Like some of the most interesting games people have ever devised, the study of probability requires only a few, very basic rules. If we use rules based on the same structure that is used for the mathematical study of sets, we can develop a theory which is simple, consistent, and easily applied to a vast number of real-life situations. We can think of probability as being a game that is played on the background consisting of all possibilities in an experiment. We call the set of all these possibilities together the **sample space**, and the individual elements of the sample space are called **outcomes**. A subset of the sample space made up of some of these outcomes is called an **event**.

Question   If a die is rolled, what is the probability of getting a five?

Answer   Since there are a total of 6 equally likely possibilities, or outcomes, in the sample space, the probability of getting one of them in particular—for instance, a five—is ⅙.

Basic Rule	If there are a total of *N* equally likely outcomes in an experiment, then the probability of any particular one of these occurring is $1/N$.

Example If a person is chosen at random from a classroom of 20 people, what is the probability of any one particular individual being chosen?

Solution According to our basic rule with $N = 20$, the probability of any one particular individual being chosen is $\frac{1}{20}$.

Question If a die is rolled, what is the probability of getting a number higher than four?

Answer Since there are 6 possible outcomes altogether, and 2 of these outcomes make up the event of getting a number higher than four, the probability of getting a number higher than four is $\frac{2}{6}$.

> **Basic Rule** If there are a total of N equally likely outcomes in an experiment, and n of these make up a particular event, then the probability of that event is $\frac{n}{N}$.

Example If there are 20 people in a room, and 12 of them are female, what is the probability that a person chosen at random from the room is a female?

Solution We can apply our basic rule, taking $N = 20$ and $n = 12$, and find the probability of choosing a female to be $\frac{12}{20}$. [Since $\frac{12}{20} = \frac{60}{100} = .60$, we can also say that there is a 60% chance of choosing a female.]

♦   ♦   ♦

Example Suppose a container has 10 blocks, of which 6 are blue and 4 are not. If 3 blocks are chosen at random, what is the probability of getting 2 that are blue and 1 that is not?

Solution In an example on page 12, we found that there are, altogether,

$$N = \binom{10}{3} = 120$$

ways to choose 3 from 10 blocks. In another example on page 12 we found that there are

$$n = \binom{6}{2} \times \binom{4}{1} = 60$$

ways to choose 2 from the 6 blue blocks and 1 from the 4 non-blue blocks. Therefore, when 3 blocks are chosen at random from 10 blocks where 6 are blue

and 4 are not, the probability of getting 2 blue and 1 not blue is $^{60}/N = {}^{60}/_{120} =$ ½.

In order for probabilities to make sense, our intuition tells us that certain basic rules must govern their behavior. The four basic rules that follow help set up the necessary foundation so that we can manipulate probabilities, just as the rules of addition and multiplication enable us to manipulate numbers.

Question   If an ordinary die is rolled once, what is the probability of rolling a seven?

Answer   It is impossible to roll a seven with one ordinary die, so the probability of this event must be zero.

Basic Rule	The probability of an impossible event is zero.

Question   If an ordinary die is rolled once, what is the probability of rolling a number between (and including) one and six?

Answer   Since the event of rolling a number between one and six includes all the possibilities, or outcomes, in this experiment, it must have a probability equal to one.

Basic Rule	The probability of the event that includes all possibilities—that is, the full sample space—in an experiment is equal to one.

Question   If an ordinary die is rolled, what is the probability that the number rolled will *not* be more than four?

Answer   We already know the probability of getting a number more than four is ⅔, so the probability of *not* getting a number more than four must be 1 - ⅔ or ⅓.

Basic Rule	The probability of an event *not* occurring is one minus the probability that it *does* occur.

Question If an ordinary die is rolled, what is the probability that the number rolled will be either less than two or more than four?

Answer The probability of the number rolled being less than two is $\frac{1}{6}$, and the probability of the number being more than four is $\frac{2}{6}$. [Notice that being less than two and being more than four are two events that cannot both occur together.] Altogether, the probability of the number being less than two *or* more than four is $\frac{1}{6} + \frac{2}{6} = \frac{3}{6}$.

Basic Rule **[Special Addition Rule]** If there are two events which cannot both occur together, then the probability of one or the other occurring is the sum of each of their probabilities.

Notation for Probability

Our intuition tells us that if something cannot happen, its probability is zero, and if it must happen, its probability is one. Also, if the probability of rain on a certain day is .70, then obviously the probability of no rain is .30. If the probability of picking a red M&M™ from a bag is .20 and the probability of picking a green one is .15, then the probability of picking a red or a green must be .35. However, there are many probability problems which are much less straightforward than these, and sometimes our intuition alone cannot lead us to a correct solution. For cases like these, formal rules can take over when our intuition is not enough. These rules can be stated more simply, and applied more systematically, if good notation is used.

We denote the full sample space of all possible outcomes in an experiment as *Fullspace*, and an event, or subset, in the full sample space is denoted *Event* or some other capitalized word (or a capital letter). If there are no outcomes in an event, then that event is an empty set, denoted *Emptyspace*. [For example, the event of rolling a seven with one die is the empty set "*Emptyspace*."]

The probability of the event "*Event*" is denoted Prob(*Event*). [For example, for "*Greater4*" the event of rolling a die and getting a number greater than four, Prob(*Greater4*) = $\frac{2}{6}$.]

The situation in which an *Event* does not occur is called the **complement** of the *Event*, denoted "not-*Event*." [For example, for the event *Greater4* that a single dice roll produces a number greater than four, the probability of the event that the number rolled is *not* greater than four can be written as Prob(not-*Greater4*) = $\frac{4}{6}$.]

This notation lets us list the previous four rules as follows.

1. Prob(*Emptyspace*) = 0.
2. Prob(*Fullspace*) = 1.

3. Prob(not-*Event*) = 1 - Prob(*Event*).
4. Prob(*Event1* or *Event2*) = Prob(*Event1*) + Prob(*Event2*) for any two events *Event1* and *Event2* which cannot occur together.

Example   If three dice are rolled, what is the probability of the event *Pair* of getting a pair (that is, two dice the same and the other different)?

Solution   We found in an example on page 3 that there are a total of *N* = 6 x 6 x 6 = 216 equally likely outcomes when three dice are rolled. From another example on page 10, we know that

$$n = 6 \times \binom{3}{2} \times 5 = 90$$

of these make up the event *Pair* of getting a pair, so we can write Prob(*Pair*) = $^{90}/_{216}$.

[Since $^{90}/_{216}$ is approximately .42, we can also say that about 42% of the time, the roll of three dice will result in a pair.]

◆   ◆   ◆

Example   If three dice are rolled, what is the probability of the event *Three* of getting three-of-a-kind?

Solution   In the example on page 11, we saw that out of the total of *N* = 216 equally likely outcomes, *n* = 6 result in all three the same number, so Prob(*Three*) = $^{6}/_{216}$, or about .03.

◆   ◆   ◆

Example   If three dice are rolled, what is the probability of the event *Either* that either a pair or three-of-a-kind are rolled?

Solution   Since we assume that the third die is different when a pair is rolled, the events of rolling a pair and three-of-a-kind cannot both occur together. Therefore, Prob(*Either*) = Prob(*Pair*) + Prob(*Three*) = $^{90}/_{216}$ + $^{6}/_{216}$ = $^{96}/_{216}$, which is approximately .44.

◆   ◆   ◆

Example   If three dice are rolled, what is the probability of the event *Different* that all three are different?

Solution   The event *Different* of getting all three dice different is the complement of the event *Either* of getting either a pair or three-of-a-kind, so we have

Prob(*Different*) = Prob(not-*Either*) = 1-Prob(*Either*) = 1 - $^{96}/_{216}$ = $^{120}/_{216}$, or about .56.

Alternate Solution   The total number *N* of possibilities when three dice are rolled is 6 x 6 x 6 = 216. According to an example on page 12, the number

n of ways to get all three dice different is 6 x 5 x 4 = 120, so Prob(*Different*) = $^{120}/_{216}$.

Experiment:
Probabilities vs. Actual Proportions

Materials: cards, dice, dominoes, roulette wheel, containers, and blocks: six blue and four not blue

1.   Probabilities
   *   What is the probability of **not** getting a face card when picking a card at random from a deck of 52? [A face card is a jack, queen, or king.]
   *   For the roll of two dice there are 6 x 6 = 36 possible outcomes. What is the probability of getting a pair of fives? [Hint: turn to page 33 if you want to see a list of all 36 outcomes.]
   *   For the roll of two dice, what is the probability of getting any pair at all?
   *   For the roll of two dice, how many ways are there altogether to get a total of seven points? What is the probability of getting seven points?
   *   There are 28 dominoes in a set, 7 of which are doubles (for blank through six). What is the probability of getting a "double" when picking one domino at random from an ordinary set?
   *   A roulette wheel has all the numbers from 1 to 36, as well as the numbers 0 and 00 which count as neither even nor odd. What is the probability of spinning an odd number on a roulette wheel?
   *   In a container are 10 blocks of which 6 are blue and 4 are not. If 3 blocks are selected at random, what is the probability that 2 are blue and 1 is not?

2.   Actual Proportions
   *   Pick a card at random from a shuffled deck; replace it, shuffle, pick again, and so forth, for a total of 13 cards. What proportion of those 13 cards are not face cards?
   *   Roll a pair of dice 36 times, recording the number and finally the proportion of times you get
      ➡   a pair of fives;
      ➡   any pair;
      ➡   a total of 7 points.
   *   Pick a domino at random from a set of 28; replace it, pick again, and so forth, for a total of eight dominoes. What proportion of them are doubles?
   *   Spin a roulette wheel 19 times. What proportion of times do you spin an odd number?

• Pick 3 blocks from a container which has 6 blue and 4 not blue. Do this 10 times. What proportion of times do you get 2 blue and 1 not blue?

3. Considering all of your results above, how do probabilities in theory relate to actual proportions occurring in the short run?

4. (Optional) If results were compiled for a class of many students rolling dice, picking cards, and so forth, would you expect a comparable relation between probabilities and actual proportions? Compile the results for your class and then compare the closeness between probabilities and actual proportions for any one class member to the closeness between probabilities and actual proportions for the entire class. Is it as you expected?

● ●

Cards: Proportion of Non-Face Cards in 13 Random Selections

Name	Proportion of Non-Face Cards
N. Pfenning	$9/13$

Dice: Rolling a Pair of Dice 36 Times, Proportion of Times You Rolled ...

Name	Pair of Fives	Any Pair	Total of 7 Points
N. Pfenning	$0/36$	$7/36$	$6/36$

● ●

Dominoes: Proportion of Doubles in 8 Random Selections

Name	Proportion of Doubles
N. Pfenning	⅛

Roulette: Proportion of Odd Numbers in 19 Spins

Name	Proportion of Odd Numbers
N. Pfenning	8/19

**Colored Blocks: Randomly Select 3 from 10,
of Which 6 Are Blue, 4 Are Not (10 Trials)**

Name	Proportion of Times 2 Blue and 1 Not Blue Chosen
N. Pfenning	$^5/_{10}$

Exercise:
Birthdays (Calculator Required)

1. Explain why, if three people are chosen at random, the probability that all three have birthdays on different days is $^{(365 \times 364 \times 363)}/_{(365 \times 365 \times 365)}$. Rewrite the above probability as $1 \times {}^{364}/_{365} \times {}^{363}/_{365}$ and solve for it using a calculator.

2. Explain why, if three people are chosen at random, the probability that at least two of them have the *same* birthday is $1 - (1 \times {}^{364}/_{365} \times {}^{363}/_{365})$.

3. Write an expression for the probability that five people chosen at random all have birthdays on *different* days. Solve for this probability using a calculator.

4. Take your solution for the previous question and subtract it from 1 to find the probability that, if five people are chosen at random, at least two of them share the *same* birthday.

5. (Optional) How many students are in your class? Find the probability that they all have *different* birthdays. Then find the probability that at least two of them share the *same* birthday.

6. (Optional) How many students must be in a class in order for the probability to be *less* than 50% that all have *different* birthdays? Note that for this same number of students, the probability is *more* than 50% that at least two of them share the *same* birthday. How many students must be in a class in order for the probability to be *more* than 90% that at least two of them share the *same* birthday?

7. (Optional) Make up (and solve) a probability problem involving dice, cards, dominoes, or roulette.

Venn Diagrams

Often the rules of probability are easier to visualize, and to apply, if we depict events in the setting of a sample space. This can be done with Venn Diagrams. A rectangle usually represents the full sample space *Fullspace*, and circles inside the rectangle represent events. If the circles do not overlap, this shows that the events cannot occur together. If the circles do overlap, then both events may occur together and the region of overlap represents the event that both occur.

Question   Out of a class of 20 students, 12 are female. Altogether, 10 students ride the bus to school. In particular, 8 females ride the bus.

    1. How many of the females do not ride the bus?
    2. How many of the students who ride the bus are male?
    3. How many students in the class are males who do not ride the bus?

Answer   On the Venn diagram, there should be 12 altogether in the female circle *Female*, and 10 altogether in the bus circle *Bus*. To depict the number of students in each category accurately, we must decide how many to place inside and outside of the overlap region, so that is the best place to begin. We know that 8 students belong in the overlap region, because they are female *and* they ride the bus.

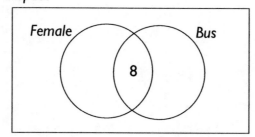

Fullspace

1. Since there are a total of 12 females, there remain 4 females to be placed in the part of the *Female* circle that is not contained in the *Bus* circle: thus, we have shown that 4 females do not ride the bus.

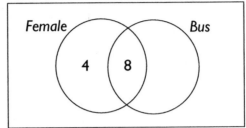

Fullspace

2. Since there are a total of 10 bus-riders, there remain 2 students to be placed in the part of the *Bus* circle that is not contained in the *Female* circle: thus, we have shown that there are 2 male students who ride the bus.

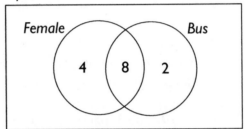

Fullspace

3. Finally, since the entire class consists of 20 students, and we have accounted for 4 + 8 + 2 = 14 of them in the *Female* and *Bus* circles, there remain 20 - 14 = 6 students who are neither female nor ride the bus: they are males who do not ride the bus.

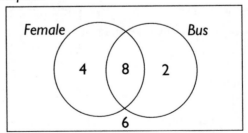

Fullspace

Question   Out of a class of students, 60% are females. Altogether, 50% of the students ride the bus. In particular, 40% of the students are females who ride the bus.

1.   If a student is picked at random from the class, what is the probability that the student is a female who does not ride the bus?
2.   What is the probability that the student is a male who *does* ride the bus?
3.   What is the probability that the student is a male who does *not* ride the bus?

Answer   Venn diagrams may also be used to depict *probabilities* of events instead of *counts* of elements in certain events. We start with a probability of .40 for a randomly chosen student being female *and* riding the bus.

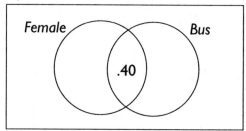

Fullspace

1.   We subtract .40 from .60 to find that .20 is the probability that a randomly chosen student is female and does not ride the bus.

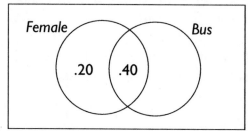

Fullspace

2. Subtracting .40 from .50 shows that .10 is the probability that a randomly chosen student is a male and rides the bus.

Fullspace

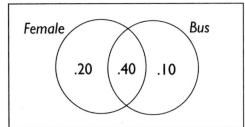

3. Altogether we must account for 100% of the students in our sample space *Fullspace*; this leaves 30% or .30 in the region outside *Female* and *Bus*. Thus, the probability is .30 that a randomly chosen student is a male who does not ride the bus.

Fullspace

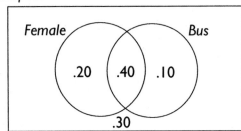

Question   As before, out of a class of students, 60% are female. Altogether, 50% of the students ride the bus. In particular, 40% of the students are females who ride the bus. If a student is chosen at random from the class, what is the probability that the student is a female *or* rides the bus (or both)?

Answer   From the Venn diagram, we see that altogether .20 + .40 + .10 = .70 is the probability of belonging in either the *Female* circle or the *Bus* circle (or both).

Alternate Answer   We can get the same answer by adding the probabilities of being a female (.60) and riding the bus (.50), but we must also subtract the probability of being a female *and* riding the bus (.40) so that the overlap region is not counted twice. Using this method we find that the probability of a randomly chosen student being a female or riding the bus is .60 + .50 - .40 = .70

Basic Rule	**[General Addition Rule]** For any two events *Event1* and *Event2* in a sample space, the probability of either (or both) occurring is the sum of each of their probabilities, minus the probability that both occur. We can write

$$\text{Prob}(Event1 \text{ or } Event2)$$
$$= \text{Prob}(Event1) + \text{Prob}(Event2) - \text{Prob}(Event1 \text{ and } Event2).$$

Example   What is the probability that the number rolled with a single die is either even or greater than four (or both)?

Solution   The probability of the event *Even* that the number rolled is even is ³⁄₆. The probability of the event *Greater4* that the number rolled is greater than four is ²⁄₆. The probability of the event that the number rolled is both even *and* greater than four is ¹⁄₆ (because this is the same as the probability of the event that a six is rolled). According to our rule, Prob(*Even* or *Greater4*) = Prob(*Even*) + Prob(*Greater4*) - Prob(*Even* and *Greater4*) = ³⁄₆ + ²⁄₆ - ¹⁄₆ = ⁴⁄₆. Thus, ⁴⁄₆ is the probability that the number rolled with a single die is even or greater than four (or both).

Alternate Solution   In fact, the total number of possible outcomes for the roll of one die is $N = 6$. The event of the number being even or greater than four is made up of $n = 4$ outcomes, namely, that the number rolled is two, four, five, or six. We can calculate the probability to be $n/N = $ ⁴⁄₆.

Independent Events

Question   If a die is rolled twice, what is the probability of rolling a five both times?

Answer   Denoting each possibility as (*first roll, second roll*), we can list all $N = 6 \times 6 = 36$ possibilities in this two-stage process as follows:

```
(1,1)  (2,1)  (3,1)  (4,1)  (5,1)  (6,1)
(1,2)  (2,2)  (3,2)  (4,2)  (5,2)  (6,2)
(1,3)  (2,3)  (3,3)  (4,3)  (5,3)  (6,3)
(1,4)  (2,4)  (3,4)  (4,4)  (5,4)  (6,4)
(1,5)  (2,5)  (3,5)  (4,5)  (5,5)  (6,5)
(1,6)  (2,6)  (3,6)  (4,6)  (5,6)  (6,6)
```

Of these, only $n = 1$ outcome makes up the event of rolling two fives: (5,5). The probability of rolling a five both times is $\frac{1}{36} = \frac{1}{36}$.

Alternate Answer	The probability of a five coming up on the first roll is $\frac{1}{6}$. Since the outcome on the second roll is "independent" of the outcome on the first roll, the probability of getting a five on the second roll is also $\frac{1}{6}$. Altogether, the probability of getting a five on the first roll, and then another five on the second roll, is $\frac{1}{6} \times \frac{1}{6} = \frac{1}{36}$.

Basic Rule   [**Special Multiplication Rule**] Suppose there are two **independent events**. That is, whether or not one occurs does not affect the probability that the other occurs. Then the probability that both occur is the product of each of their probabilities: Prob(*Event1* and *Event2*) = Prob(*Event1*) x Prob(*Event2*).

Example	If a card is chosen at random from an ordinary deck of 52 and replaced and then another card is chosen, what is the probability that both are red cards?
Solution	The probability of the event *First* that the first card chosen is red is $\frac{26}{52} = \frac{1}{2}$. The probability of the event *Second* that the second card is red is also $\frac{26}{52} = \frac{1}{2}$, since the first card has been replaced. According to our rule, the probability that both cards are red is Prob(*First* and *Second*) = Prob(*First*) x Prob(*Second*) = $\frac{1}{2} \times \frac{1}{2} = \frac{1}{4}$.

Conditional Probabilities

Question	In a class of 20 students, 12 are female. Altogether, 10 students ride the bus to school. Specifically, 8 females ride the bus to school. The probability of the event *Female* that a randomly chosen student is female is $\frac{12}{20} = .60$, and the probability of the event *Bus* that a randomly chosen student rides the bus is $\frac{10}{20} = .50$. Can we use our multiplication rule above to find the probability that a randomly chosen student is female *and* rides the bus?
Answer	If we knew that the events *Female* and *Bus* were independent, we could say that Prob(*Female* and *Bus*) would be equal to Prob(*Female*) x Prob(*Bus*) = .60 x .50 = .30. However, whether or not a student is female may affect the probability that the student rides the bus, and independence may not be taken for granted. The probability that a randomly chosen student is female and rides the bus is *not* necessarily .30.

[Note that we are considering here the issue of independence *within this class of 20 students* only. It may be that in a population of thousands of students, whether or not a student takes the bus is roughly independent of gender. If that were so, then the probability of being female and riding the bus would approximately equal the product of the probabilities of being female and of riding the bus within that larger group.]

We have said that for independent events, whether or not one occurs does not affect the probability of the other occurring. We will call two events **dependent** if whether or not one occurs may affect the probability of the other occurring. In order to develop a rule which works for events that are dependent, it may help to begin by looking at a "two-way table," where elements of *Fullspace* are counted as being in or out of either of two categories. For the class of 20 students, we have the following two-way table:

	Female	Male	Total
Rides Bus	8	2	10
Does Not Ride Bus	4	6	10
Total	12	8	20

Question According to the two-way table above, what is the probability that a randomly chosen student is female and rides the bus?

Answer Altogether there are $N = 20$ students in the class. Out of these, $n = 8$ make up the event of being female and riding the bus. Therefore, the probability is $n/N = 8/20 = .40$ that a randomly chosen student is female and rides the bus. In this case, Prob(*Female* and *Bus*) does not equal Prob(*Female*) x Prob(*Bus*), since .40 does not equal .30.

The sample space *Fullspace* may change within a problem. If a certain condition is given, then we must solve for probabilities based on that condition. The concept of conditional probabilities enables us to analyze situations where the sample space changes. Thus, conditional probabilities will enable us to state a rule for the probability of two *dependent* events both occurring.

Question If a student is chosen at random *from the females* of the class represented in the two-way table above, what is the probability that she rides the bus?

Answer Since the student is chosen from the females, the sample space *Fullspace* now has only 12 elements. In our two-way table above, we now restrict our attention to the Female column only. [If we wanted to refer to our Venn diagram on page 32 instead, we would

ignore everything except the Female circle, which would now represent the entire sample space.] Out of these 12 females, there are $n = 8$ who ride the bus. The probability of riding the bus, *given* that a student is female, is $n/N = 8/12$.

In the question above, we were first given the *condition* that the student be female. In general, the **conditional probability** of *Event2* given *Event1*, written Prob(*Event2* given *Event1*), is the probability that *Event2* occurs given that *Event1* occurs. Thus, in this particular question, we are asked to find Prob(*Bus* given *Female*), that is, the probability of riding the bus given that a student is female. Using this notation, our answer can be written as Prob(*Bus* given *Female*) = $8/12$.

Basic Rule	For any two events *Event1* and *Event2*, Prob(*Event2* given *Event1*) equals the number of outcomes in *Event2* and *Event1*, divided by the number of outcomes in *Event1*.

Example What is the probability that the number rolled with one die is even given that it is less than four?

Solution There is only one outcome in *Even* and *Less4* [that is, the event of rolling a two] and there are three outcomes in *Less4* [that is, the event of rolling a one, two, or three]. Therefore, the probability that the number rolled with one die is even given that it is less than four is Prob(*Even* given *Less4*) = $1/3$.

◆   ◆   ◆

Example Use the two-way table on page 35 to solve for the following:

1. The probability of riding the bus given that a student is male;
2. The probability of being female given that a student rides the bus;
3. The probability of being male given that a student rides the bus.

Solution
1. Prob(*Bus* given *Male*) = $2/8$.
2. Prob(*Female* given *Bus*) = $8/10$.
3. Prob(*Male* given *Bus*) = $2/10$.

Question For the same class represented in the two-way table on page 35, can we say that the event of riding the bus is independent of whether or not a student is female?

Answer The overall probability of riding the bus is Prob(*Bus*) = $10/20 = 1/2$. But the probability of riding the bus, given that a student is female, is

Prob(*Bus* given *Female*) = ⁸⁄₁₂ = ⅔. Whether or not a student is female affects the probability that the student will ride the bus. In this class, females happen to be more likely to ride the bus. Therefore, the events are **not** independent.

This kind of reasoning provides us with a simple and reliable rule for determining whether or not two events are independent.

Basic Rule	*Event*2 is independent of *Event*1 if Prob(*Event*2) = Prob(*Event*2 given *Event*1). In other words, *Event*2 is independent of *Event*1 if the probability of *Event*2 occurring is not affected by the occurrence of *Event*1.

Example   For a single dice roll, is the event *Even* of getting an even number independent of the event *Less*4 that the number rolled is less than four?

Solution   Prob(*Even*) = ½, but Prob(*Even* given *Less*4) = ⅓. Since Prob(*Even*) does not equal Prob(*Even* given *Less*4), the events are not independent.

[With no condition given, the probability of getting an even number on one dice roll is Prob(*Even*) = ³⁄₆ = ½. But if we are given the condition that the number be less than four—namely, one, two, or three—the probability of getting an even number is only Prob(*Even* given *Less*4) = ⅓.]

◆   ◆   ◆

Example   For a single dice roll, is the event *Even* of getting an even number independent of the event *Less*5 that the number rolled is less than five?

Solution   Prob(*Even*) = ³⁄₆ = ½ and Prob(*Even* given *Less*5) = ²⁄₄ = ½. Since Prob(*Even*) = Prob(*Even* given *Less*5), the events are independent.

[The condition of being less than five does not affect the probability of getting an even number, because half of the numbers are still even. Thus, the events are independent.]

Experiment:
Venn Diagrams

Materials: completed surveys

Use Venn Diagrams to examine the relationships among and between various events.

1. With the information from class surveys, fill in the correct number of students in each section of a Venn Diagram which consists of three circles. [Use the results below if a survey for your own class is not available.]

Survey #	Male?	Has Siblings?	Has Own Room?
1	Y	Y	
2		Y	
3			Y
4		Y	Y
5	Y	Y	Y
6			Y
7	Y	Y	Y
8	Y	Y	
9	Y	Y	Y
10	Y	Y	Y
11	Y		Y
12	Y	Y	Y
13	Y	Y	Y
14	Y	Y	Y
15		Y	Y

2. Compare Prob(*Room*) to Prob(*Room* given *Siblings*) to determine if, in this particular class, having a room to one's self is independent of whether or not someone has siblings. If not, is a class member more likely to have his or her own room if he or she has siblings or if he or she is an only child? [Remember that Prob(*Room*) is just the number of members in *Room* divided by the total number of members. Prob(*Room* given *Siblings*) is the number of members in both *Room* and *Siblings* divided by the number of members in *Siblings*.]

3. Next, compare Prob(*Room*) to Prob(*Room* given *Male*) to determine if, in this particular class, having a room to one's self is independent of whether someone is male or female.

4. If all American students in your age bracket were surveyed, would you expect *Room* and *Siblings* to be independent (approximately)? Explain.

5. If all American students in your age bracket were surveyed, would you expect *Room* and *Male* to be independent (approximately)? Explain.

Exercise: Venn Diagrams

1. Pick any two columns from this [or your own] class survey, and use the information from those columns to make a Venn Diagram: Use a capital letter or a capitalized word to label each of the two circles in your diagram and fill in the number of students belonging in each region. Then make a two-way table to represent the information in your Venn Diagram.

2. Pick any three columns, and use the information from those columns to make a Venn Diagram, with the correct number of students in each region. You will need three circles for this diagram.

3. (Optional) Would four circles be enough to construct a Venn Diagram for information from four columns? Explain.

4. (Optional) Ask friends, family, or classmates to fill out a survey about some topics that interest you. Make up (and solve) a Venn Diagram problem based on your survey results.

Survey #	Male?	Siblings?	Pet?	Own Room?	Takes Bus?	Likes Math Best?
1	Y	Y			Y	Y
2		Y	Y		Y	Y
3				Y	Y	Y
4		Y	Y	Y	Y	Y
5	Y	Y		Y		
6			Y	Y	Y	Y
7	Y	Y	Y	Y	Y	Y
8	Y	Y				Y
9	Y	Y	Y	Y	Y	
10	Y	Y	Y	Y		
11	Y		Y	Y		Y
12	Y	Y		Y	Y	Y
13	Y	Y	Y	Y		Y
14		Y	Y	Y		Y
15		Y		Y		Y

More Conditional Probabilities:
Events That Occur In Stages

Question   A card is chosen at random from a set of 4 cards, of which 2 are red and 2 are black. Another is chosen *without* replacing the first. What is the probability that both are red cards?

Answer   Altogether there are

$$N = \binom{4}{2}$$

ways to choose 2 cards from a set of 4, and

$$n = \binom{2}{2}$$

ways to choose 2 out of the 2 red cards. Therefore, the probability of choosing 2 red cards is

$$n/N = \frac{\binom{2}{2}}{\binom{4}{2}} = \frac{1}{(4 \times 3)/(2 \times 1)} = 1/6$$

Alternate Answer   Let *First* be the event that the first card is red, and *Second* be the event that the second card is red. The probability Prob(*First*) of the first card being red is $2/4 = 1/2$. The probability Prob(*Second* given *First*) of the second card being red, given that the first was red, is $1/3$ because there are three cards left to choose from, of which one is red. Therefore, the probability that both are red is Prob(*First and Second*) = Prob(*First*) x Prob(*Second* given *First*) = $1/2$ x $1/3$ = $1/6$.

◆   ◆   ◆

Question   A card is chosen at random from an ordinary deck of 52 and then another is chosen *without* replacing the first. What is the probability that both are red cards?

Answer   Altogether there are

$$N = \binom{52}{2}$$

ways to choose 2 cards from a deck of 52, and

$$n = \binom{26}{2}$$

ways to choose 2 red cards out of the 26 red cards in a deck. Therefore, the probability of choosing 2 red cards is

$$\frac{n}{N} = \frac{\binom{26}{2}}{\binom{52}{2}} = \frac{(26 \times 25)/(2 \times 1)}{(52 \times 51)/(2 \times 1)} = \frac{(26 \times 25)}{(52 \times 51)} = \frac{25}{102}.$$

Alternate Answer   Let *First* be the event that the first card is red, and *Second* be the event that the second card is red. The probability Prob(*First*) of the first card being red is $\frac{26}{52} = \frac{1}{2}$. The probability Prob(*Second* given *First*) of the second card being red given that the first was red is $\frac{25}{51}$ because there are 51 cards left to choose from, of which 25 are red. Therefore, the probability that both are red is Prob(*First* and *Second*) = Prob(*First*) x Prob(*Second* given *First*) = $\frac{1}{2} \times \frac{25}{51} = \frac{25}{102}$.

Basic Rule   **[General Multiplication Rule]** The probability of two events both occurring is the product of the probability that the first occurs and the probability that the second occurs given that the first occurred:

Prob(*Event1* and *Event2*) = Prob(*Event1*) x Prob(*Event2* given *Event1*).

Example   The probability that a sunflower seed from a certain packet will sprout is 50%. Given that a seed has sprouted, the probability that it grows to produce a flower is 40%. If a seed is chosen at random from the packet, what is the probability that it will sprout *and* produce a flower?

Solution   If we let *Sprout* denote the event that a randomly chosen seed sprouts, and *Flower* the event that a randomly chosen seed produces a flower, we know that Prob(*Sprout*) = 50% and Prob(*Flower* given *Sprout*) = 40%. We want to find the probability Prob(*Sprout* and *Flower*) that a seed sprouts and produces a flower.

According to our General Multiplication Rule,
Prob(*Sprout* and *Flower*)
= Prob(*Sprout*) x Prob(*Flower* given *Sprout*) = .50 x .40 = .20.

Example   If two doughnuts are chosen at random from a bag containing three jelly and five cream doughnuts, what is the probability of getting two cream doughnuts?

Solution   The probability of the first doughnut having cream is $\frac{5}{8}$. Given that the first doughnut has cream, the probability of the second doughnut having cream is $\frac{4}{7}$, because now four of the seven remaining doughnuts have cream. Therefore, the probability of getting two cream doughnuts is $\frac{5}{8} \times \frac{4}{7} = \frac{20}{56}$.

Question   If three cards are chosen in succession (that is, without replacement) from a deck of 52, what is the probability that the first two are red and the third is black?

Answer   We found that the probability of the first two cards being red is $\frac{26}{52} \times \frac{25}{51}$. The probability that the third card is black, given that the first two cards are red, is $\frac{26}{50}$. Therefore, the probability of two red cards followed by a black card is $\frac{26}{52} \times \frac{25}{51} \times \frac{26}{50} = \frac{13}{102}$. [Because we must be specific about the order of red cards and black cards chosen, this problem does not lend itself to an alternate solution with combinations.]

In general, we have
$$\text{Prob}(Event1 \text{ and } Event2 \text{ and } Event3)$$
$$= \text{Prob}(Event1)$$
$$\times \text{Prob}(Event2 \text{ given } Event1)$$
$$\times \text{Prob}(Event3 \text{ given } (Event1 \text{ and } Event2)).$$

Example   If three students are chosen at random from 12, of whom eight are boys and four are girls, what is the probability that two boys and then one girl, in that order, are chosen?

Solution   We multiply the probabilities that the first is a boy, that the second is a boy given that the first was a boy, and that the third is a girl given that the first two were boys: $\frac{8}{12} \times \frac{7}{11} \times \frac{4}{10} = \frac{28}{165}$.

Conditional probabilities are useful when events happen in stages. In the first stage, either *Event1* occurs or not-*Event1* occurs. In the second stage, either *Event2* occurs or not-*Event2* occurs (see the tree diagram on the next page). If *Event1* and *Event2* are dependent, then the probability of the event *Event2* may vary, depending on whether or not *Event1* occurred.

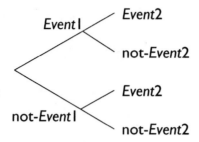

Question   The probability that Vanessa watches television while doing her homework on any given day is .65. The probability of finishing her homework by the end of the day is .60 if she watches television; it is .80 if she does not watch television. On any given day, what is the probability of Vanessa finishing her homework?

Answer   Many students prefer to solve such problems with tree diagrams, which give a visual sense to the application of the rules of probability. If we use *TV* to denote the event of watching television, and *Finish* the event of finishing homework, we know that Prob(*TV*) = .65 so the probability of not watching television is Prob(not-*TV*) = 1 - .65 = .35. We also know that Prob(*Finish* given *TV*) = .60 and Prob(*Finish* given not-*TV*) = .80:

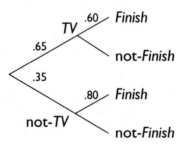

For one event *and* then another occurring [following a single branch segment and then the next], probabilities must be *multiplied*. For one sequence of events *or* another occurring [following connected branches to one end leaf or other connected branches to another end leaf], probabilities must be *added*.

The probability of reaching a *Finish* branch by way of a *TV* branch is found by multiplying .65 x .60 = .39. The probability of reaching a *Finish* branch by way of a not-*TV* branch is found by multiplying .35 x .80 = .28. Overall the probability of reaching *Finish* in the end is found by adding the probabilities of following one or the other route: .39 + .28 = .67.

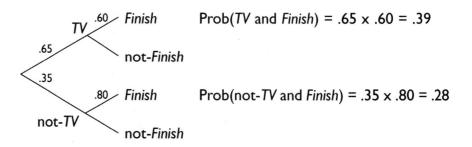

Prob(*TV* and *Finish*) = .65 x .60 = .39

Prob(not-*TV* and *Finish*) = .35 x .80 = .28

Verifying that the probabilities for all possible outcomes in Vanessa's homework "experiment" add up to one is a good way to check that the tree diagram has been filled in correctly. Below we include probabilities for Vanessa *not* finishing her homework, and see that all four probabilities together add up to .39 + .26 + .28 + .07 = 1.

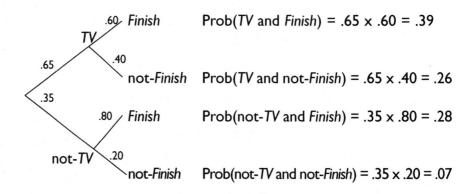

Prob(*TV* and *Finish*) = .65 x .60 = .39

Prob(*TV* and not-*Finish*) = .65 x .40 = .26

Prob(not-*TV* and *Finish*) = .35 x .80 = .28

Prob(not-*TV* and not-*Finish*) = .35 x .20 = .07

Alternate Answer

The end result of Vanessa finishing her homework may be accomplished in two ways: either she watches television and finishes her homework, or she does not watch television and finishes her homework. We can write *Finish* = (*TV* and *Finish*) or (not-*TV* and *Finish*), as shown in the Venn Diagram.

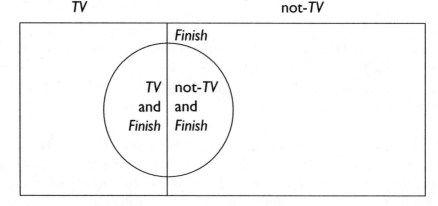

Since (*TV* and *Finish*) and (not-*TV* and *Finish*) cannot occur together, we can find the probability of one or the other occurring by adding their probabilities (General Addition Rule on page 33). Then we deal with Prob(*TV* and *Finish*) and Prob(not-*TV* and *Finish*) by using our General Multiplication Rule on page 41:

Prob(*Finish*) = Prob((*TV* and *Finish*) or (not-*TV* and *Finish*))

= Prob(*TV* and *Finish*) + Prob(not-*TV* and *Finish*)

= (Prob(*TV*) x Prob(*Finish* given *TV*)) + (Prob(not-*TV*) x Prob(*Finish* given not-*TV*))

= (.65 x .60) + (.35 x .80) = .39 + .28 = .67.

[Notice that the ordinary average of the two probabilities of Vanessa finishing her homework would be $^{(.60 + .80)}\!/_2$ = .70. Since she watches television more than half of the time, and she is less likely to finish her homework when she watches, the overall probability of finishing her homework is somewhat lower than .70.]

Basic Rule   If *Event2* may be reached by way of *Event1* or not-*Event1*, then

Prob(*Event2*) = Prob(*Event1*) x Prob(*Event2* given *Event1*)

+ Prob(not-*Event1*) x Prob(*Event2* given not-*Event1*).

Example   When Nils is shooting a penalty kick in soccer, the probability that he shoots to the right is 80%; otherwise, he shoots to the left. When he shoots to the right, the probability of scoring is 75%. When he shoots to the left, the probability of scoring is only 15%. What is the overall probability of Nils scoring on a penalty kick?

Solution   The event *Score* of scoring on a penalty kick may be reached by way of the event *Right* of shooting to the right, or by event not-*Right* of not shooting to the right (that is, shooting to the left).

We know Prob(*Right*) = .80, so Prob(not-*Right*) = .20.

We also know that the probability of scoring on a shot to the right is Prob(*Score* given *Right*) = .75, and that the probability of scoring on a shot to the left is Prob(*Score* given not-*Right*) = .15. Thus, our basic rule tells us that the overall probability of scoring is Prob(*Score*) = Prob(*Right* and *Score*) + Prob(not *Right* and *Score*) = Prob(*Right*) x Prob(*Score* given *Right*) + Prob(not-*Right*) x Prob(*Score* given not-*Right*) = (.80 x .75) + (.20 x .15) = .60 + .03 = .63.

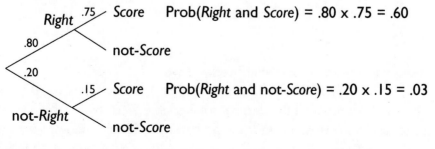

[Notice that the ordinary average of the two probabilities of scoring would be $(.75 + .15)/2$ = .45. Since Nils is more successful when he shoots to the right, and shoots to the right more often, his overall chance of scoring (.63) is quite a bit higher than .45.]

Bayes' Theorem (Optional)

We have seen that conditional probabilities can be used to tell us the probability of a particular outcome occurring at the second stage, taking into account more than one possible outcome at the first stage.

Sometimes we already know the outcome at the second stage and we want to know the probability of a particular outcome having occurred at the first stage. Such problems, because of the "backward reasoning" involved, may seem complicated at first, but they are interesting and possible as long as our Basic Rules are applied correctly.

Question   As before, we know that the probability of Vanessa watching television while doing homework is .65. The probability of finishing her homework is .60 if she watches television, and .80 if she does not watch television. Suppose we know Vanessa *did* finish her homework. What is the probability that she was watching television that day?

Answer   We want to know the probability that the event *TV* had occurred (Vanessa watched television) given that the event *Finish* occurred (Vanessa finished her homework). In the language of probability, we are looking for Prob(*TV* given *Finish*).

Since Prob(*TV* and *Finish*) = Prob(*Finish*) x Prob(*TV* given *Finish*), it must also be true that
$$\text{Prob}(TV \text{ given } Finish) = \frac{\text{Prob }(TV \text{ and } Finish)}{\text{Prob}(Finish)}.$$

From part of our answer to a previous question, we know the probability that Vanessa watches television *and* finishes her homework:

Prob(*TV* and *Finish*) = Prob(*TV*) x Prob(*Finish* given *TV*) = .65 x .60 = .39.

We also found that the overall probability of Vanessa finishing her homework is Prob(*Finish*) = .67.

Substituting into the equation above, we have
$$\text{Prob}(TV \text{ given } Finish) = \frac{\text{Prob }(TV \text{ and } Finish)}{\text{Prob}(Finish)} = \frac{.39}{.67} = .58,$$
where we have rounded the decimal solution for $\frac{.39}{.67}$ to two places. In other words, if Vanessa *did* finish her homework, the probability is .58 that she was watching television.

Looking back at the tree diagram, we see that the *Finish* branches are reached with a total probability of .39 + .28 = .67, and .39 out of the .67, or $^{.39}/_{.67}$ = .58, is the probability of reaching a *Finish* branch by way of a *TV* branch.

Basic Rule **[Bayes' Theorem]** If event *Event2* may be reached by way of event *Event1* or event not-*Event1*, then

$$\text{Prob}(\textit{Event1 given Event2}) = \frac{\text{Prob}(\textit{Event1 and Event2})}{\text{Prob}(\textit{Event2})}$$

$$=$$

$$\frac{\text{Prob}(\textit{Event1}) \times \text{Prob}(\textit{Event2 given Event1})}{\text{Prob}(\textit{Event1}) \times \text{Prob}(\textit{Event2 given Event1}) + \text{Prob}(\textit{not-Event1}) \times \text{Prob}(\textit{Event2 given not-Event1})}$$

Example A fan of Nils saw him score on a penalty kick, but couldn't remember afterwards if he shot to the right or to the left. Given that Nils scored, what is the probability that he kicked to the right?

Solution According to our Basic Rule, the probability of Nils having kicked to the right, given that he scored, is

$$\text{Prob}(\textit{Right given Score}) = \frac{\text{Prob}(\textit{Right and Score})}{\text{Prob}(\textit{Score})}$$

$$=$$

$$\frac{\text{Prob}(\textit{Right}) \times \text{Prob}(\textit{Score given Right})}{\text{Prob}(\textit{Right}) \times \text{Prob}(\textit{Score given Right}) + \text{Prob}(\textit{not-Right}) \times \text{Prob}(\textit{Score given not-Right})}$$

$$= {}^{(.80 \times .75)}/_{(.80 \times .75 + .20 \times .15)}$$

$$= {}^{.60}/_{.63} = .95$$

where we have rounded our solution to two decimal places. If Nils scored, the probability is .95 that he kicked to the right.

Notice that on the tree diagram, the *Score* branches are reached with a total probability of .60 + .03 = .63, and .60 out of the .63, or $^{.60}/_{.63}$ = .95, is the probability of reaching a *Score* branch by way of a *Right* branch.

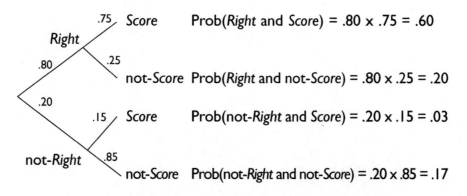

47

Also notice that the leaves in our tree diagram above take into account all four possible outcomes in the soccer "experiment," so the probabilities must total to one: .60 + .20 + .03 + .17 = 1.

However, the total may sometimes be slightly under or over one because of rounding decimals down or up.

Experiment: Labyrinth

Imagine that you are hiding a treasure inside a labyrinth. Those who are seeking the treasure, when faced with a choice of two or more possible paths, will choose any one of them with equal probability. [That is, if a path divides in 2, each branch has probability ½. If a path divides in 3, each branch has probability ⅓.] In which of the two chambers, A or B, is the treasure less likely to be found?

Hint: Chamber B can be reached by turning left **and** then right
or right **and** then left
or right **and** then right **and** then left.

Remember that the word **and** in probability indicates multiplication; the word **or** in probability indicates addition.

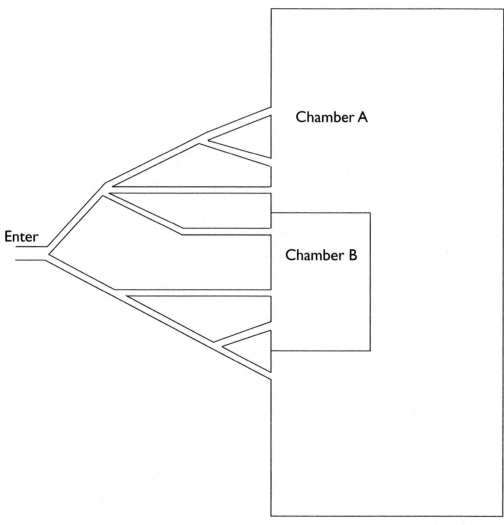

Exercise: Labyrinth

1. Imagine that you are hiding a treasure inside a labyrinth. Those who are seeking the treasure, when faced with a choice of two or more possible paths, will choose any one of them with equal probability. In which of the two chambers, A or B, is the treasure less likely to be found?

2. (Optional) Make up your own labyrinth—as difficult as you want (as long as you can solve it yourself)! Now see if your classmates can solve it.

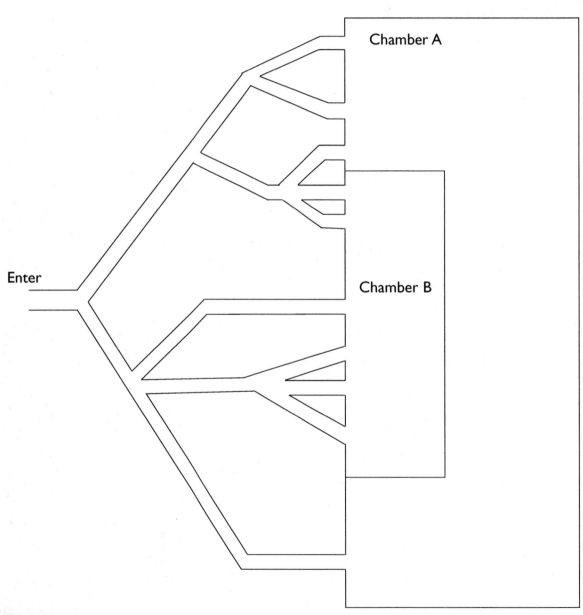

Probability Distributions

Introduction

A variable may be something that takes on various *quantities,* such as a number from one to six. Such variables are called **numerical**. Otherwise, a variable takes on various *qualities,* such as male or female. These variables are called **categorical**. [Some texts use the term **quantitative** for numerical variables and **qualitative** for categorical variables.] A categorical variable gives rise to a numerical variable when we work with the number or proportion of times the variable falls into a particular category. For example, if two students are selected from a class, the number or proportion of females selected would be a numerical variable.

The numerical variables for the number rolled on a die and for the number of females picked when students are chosen at random from a group are called **random variables** because chance alone determines what their values will be. However, in the long run, we know there will be a pattern for how often the random variable takes on each possible value. By looking at the **probability distribution** for a random variable, we can see the pattern for its behavior. In fact, **probability** as a science is the formal study of random behavior.

If we can count out all the possible values of a random variable such as the number rolled on a die or the number of tries until a six is rolled, then the random variable is **discrete**. A random variable such as one for height or weight or for an amount of time elapsed or a distance traveled has an infinite number of possible values over a continuous range. These random variables are called **continuous**. For now, we will consider the probability distributions of discrete random variables; in another chapter, we will modify these ideas to study the behavior of continuous random variables.

Probability Distributions for Independent Events

Question What are all the possible outcomes when a coin is flipped twice?

Answer The possibilities are two tails, a tail followed by a head, a head followed by a tail, and two heads.

Question   What are the probabilities for each of the four possible outcomes when a coin is flipped twice?

Answer   Each of the four outcomes is equally likely, so each has probability ¼.

Alternate Answer   On the first flip, the probability of heads is ½, and on the second flip, the probability of heads is also ½. Since the flips are independent, the probability of heads both times is ½ x ½ = ¼. In the same way, the probabilities for the other three possible outcomes can also be calculated to be ¼.

◆   ◆   ◆

Question   What values can the random variable for the number of heads in two coin flips take on, and what are the probabilities of taking on each of these values?

Answer   The number of heads can be zero, one, or two. The probability of zero heads is ¼, because only one of the four equally likely outcomes makes up this event. The probability of one head is ¼ + ¼ = ½, because two of the four equally likely outcomes make up this event. The probability of two heads is ¼, because one of the four equally likely outcomes makes up this event.

Basic Rule   If the probability distribution of a discrete random variable is not given explicitly, we may find it by looking first at the associated sample space and listing the probabilities of all possible outcomes, as well as the value of the random variable for each of these outcomes. From this, an ordered list of possible values of the random variable may be made, along with the probabilities of taking on each value. This shows the probability distribution of the random variable.

Example   Find the probability distribution of the number of heads in two coin flips.

Solution   The elements of the *Fullspace*, associated probabilities, and values of the random variable can be listed as follows, using *T* for tails and *H* for heads:

Fullspace	Probability	Number of Heads
TT	¼	0
TH	¼	1
HT	¼	1
HH	¼	2

From this we can show the probability distribution to be:

Number of Heads	Probability
0	$\frac{1}{4}$
1	$\frac{1}{2}$
2	$\frac{1}{4}$

◆ ◆ ◆

Example Find the probability distribution of the number of diamonds chosen if a card is picked from an ordinary deck and replaced, after which another card is picked and replaced, and finally a third card is picked and replaced.

Solution Because each card is replaced, we are working with independent events. Getting a diamond each time has probability $\frac{1}{4}$, whereas not getting a diamond has a probability of $\frac{3}{4}$. Using D for the event of getting a diamond, and N for the event of not getting a diamond, a tree diagram shows that there are eight possibilities to be listed in the *Fullspace*:

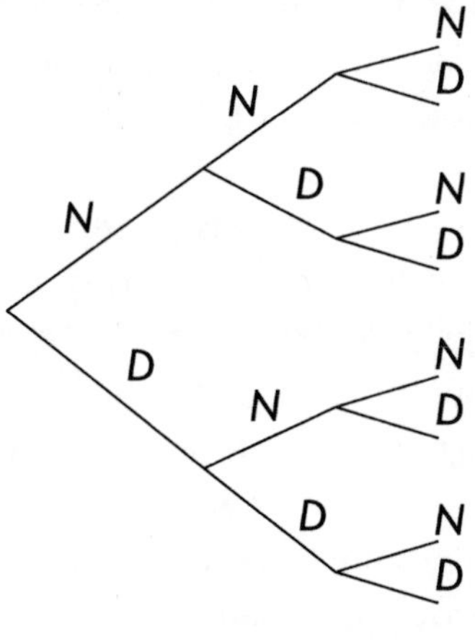

Fullspace	Probability	Number of Diamonds
NNN	$\frac{3}{4} \times \frac{3}{4} \times \frac{3}{4} = \frac{27}{64}$	0
NND	$\frac{3}{4} \times \frac{3}{4} \times \frac{1}{4} = \frac{9}{64}$	1
NDN	$\frac{3}{4} \times \frac{1}{4} \times \frac{3}{4} = \frac{9}{64}$	1
DNN	$\frac{1}{4} \times \frac{3}{4} \times \frac{3}{4} = \frac{9}{64}$	1
NDD	$\frac{3}{4} \times \frac{1}{4} \times \frac{1}{4} = \frac{3}{64}$	2
DND	$\frac{1}{4} \times \frac{3}{4} \times \frac{1}{4} = \frac{3}{64}$	2
DDN	$\frac{1}{4} \times \frac{1}{4} \times \frac{3}{4} = \frac{3}{64}$	2
DDD	$\frac{1}{4} \times \frac{1}{4} \times \frac{1}{4} = \frac{1}{64}$	3

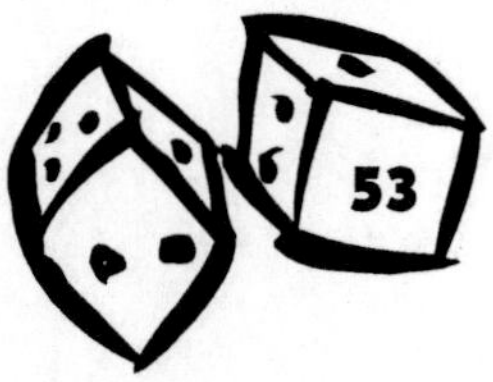

The probability distribution is:

Number of Diamonds	Probability
0	$^{27}/_{64}$
1	$^{9}/_{64} + ^{9}/_{64} + ^{9}/_{64} = ^{27}/_{64}$
2	$^{3}/_{64} + ^{3}/_{64} + ^{3}/_{64} = ^{9}/_{64}$
3	$^{1}/_{64}$

Probability Distributions for Dependent Events

Question   If two doughnuts are chosen at random from a bag containing three filled with jelly and five filled with cream, find the probability distribution of the number of jelly doughnuts chosen.

Answer   When solving for probabilities of individual outcomes, we must keep in mind that the first doughnut is not replaced before the second is chosen. [In fact, it is probably eaten immediately.] If the first doughnut has jelly, then it is less likely that the second doughnut has jelly because the number of jelly doughnuts to choose from has been reduced: the second selection is *not* independent of the first.

The sample space for this experiment contains the four elements *CC, CJ, JC,* and *JJ,* where *C* denotes a cream doughnut chosen and *J* denotes a jelly doughnut chosen. To find the probability associated with the outcome *CC,* we note that the probability of getting a cream doughnut on the first selection is $^{5}/_{8}$. [There are a total of eight doughnuts, five of which have cream.] The probability that the second has cream, given that the first has cream, is $^{4}/_{7}$. [Now there are seven doughnuts left, four of which have cream.] We multiply these and find the probability of two cream doughnuts to be $^{5}/_{8} \times ^{4}/_{7} = ^{20}/_{56}$. For this outcome, the random variable for the number of jelly doughnuts chosen is zero.

Next, the probability of getting first a cream doughnut and then a jelly doughnut (event *CJ*) is $^{5}/_{8} \times ^{3}/_{7} = ^{15}/_{56}$, and the random variable in this case equals one. Similarly, the probability of the outcome *JC* is $^{3}/_{8} \times ^{5}/_{7} = ^{15}/_{56}$ and the random variable again equals one. Finally, the probability of the outcome *JJ* is $^{3}/_{8} \times ^{2}/_{7} = ^{6}/_{56}$, and for this outcome the random variable takes the value two. We can summarize all this information as follows:

Fullspace	Probability	Number of Jelly Doughnuts
CC	$\frac{5}{8} \times \frac{4}{7} = \frac{20}{56}$	0
CJ	$\frac{5}{8} \times \frac{3}{7} = \frac{15}{56}$	1
JC	$\frac{3}{8} \times \frac{5}{7} = \frac{15}{56}$	1
JJ	$\frac{3}{8} \times \frac{2}{7} = \frac{6}{56}$	2

Number of Jelly Doughnuts	Probability
0	$\frac{20}{56} = \frac{10}{28}$
1	$\frac{15}{56} + \frac{15}{56} = \frac{15}{28}$
2	$\frac{6}{56} = \frac{3}{28}$

Alternate Answer

The above solution made use of our rule for the probability of the occurrence of two dependent events, which involves multiplying by a conditional probability. Another approach would be to solve for each of the probabilities by finding n—the number of possible ways to arrive at a certain outcome—and dividing by N—the total number of ways to perform the experiment of choosing two doughnuts.

The number of ways to choose 0 of the 3 jelly doughnuts and 2 of the 5 cream doughnuts is

$$n = \binom{3}{0} \times \binom{5}{2} = 10.$$

The total number of ways to choose 2 from 8 doughnuts is

$$N = \binom{8}{2} = 28.$$

Thus, the probability of choosing 2 cream doughnuts is $\frac{n}{N} = \frac{10}{28}$, the same as our solution above.

The number of ways to choose 1 cream and 1 jelly doughnut (in either order) is

$$\binom{5}{1} \times \binom{3}{1} = 15$$

Dividing by the total number of ways "28" gives a probability of $\frac{15}{28}$. This solution combines the probabilities of the outcomes CJ and JC, which specified the order of the doughnuts chosen.

Finally, the probability of choosing 2 of the 3 jelly doughnuts and 0 of the 5 cream doughnuts when 2 doughnuts are chosen at random from 8 is

$$\frac{\binom{3}{2} \times \binom{5}{0}}{\binom{8}{2}} = \frac{3}{28}$$

the same as our solution on the previous page.

The probability distribution based on this solution can be written in terms of combinations:

Number of Jelly Doughnuts	Probability
0	$\dfrac{\binom{3}{0} \times \binom{5}{2}}{\binom{8}{2}} = \frac{10}{28}$
1	$\dfrac{\binom{3}{1} \times \binom{5}{1}}{\binom{8}{2}} = \frac{15}{28}$
2	$\dfrac{\binom{3}{2} \times \binom{5}{0}}{\binom{8}{2}} = \frac{3}{28}$

Basic Rule   The events in a sampling-without-replacement problem are dependent. The probability distributions for the random variables in such problems may be found either by multiplying conditional probabilities or by dividing combinations.

Example   If 3 students are chosen at random from 12, of whom 8 are boys and 4 are girls, what is the probability distribution of the number of boys chosen?

Solution   We can follow the same process as in our alternate solution above, simply changing the numbers when necessary. First, the total number of ways to select 3 from 12 students is

$$\binom{12}{3} = \frac{12 \times 11 \times 10}{3 \times 2 \times 1} = 220$$

The number of ways to select 0 boys and 3 girls is

$$\binom{8}{0} \times \binom{4}{3}$$

The number of ways to select 1 boy and 2 girls is

$$\binom{8}{1} \times \binom{4}{2}$$

The number of ways to select 2 boys and 1 girl is

$$\binom{8}{2} \times \binom{4}{1}$$

The number of ways to select 3 boys and 0 girls is

$$\binom{8}{3} \times \binom{4}{0}$$

The probability distribution is

Number of Boys	Probability
0	$$\dfrac{\binom{8}{0} \times \binom{4}{3}}{\binom{12}{3}} = \frac{(1 \times 4)}{220} = \frac{1}{55}$$
1	$$\dfrac{\binom{8}{1} \times \binom{4}{2}}{\binom{12}{3}} = \frac{(8 \times 6)}{220} = \frac{12}{55}$$
2	$$\dfrac{\binom{8}{2} \times \binom{4}{1}}{\binom{12}{3}} = \frac{(28 \times 4)}{220} = \frac{28}{55}$$
3	$$\dfrac{\binom{8}{3} \times \binom{4}{0}}{\binom{12}{3}} = \frac{(56 \times 1)}{220} = \frac{14}{55}$$

[Note: If we wanted to solve this using conditional probabilities, we would have to take into account that the boys and girls can be chosen in different orders. For instance, the probability of choosing two boys and one girl could be found by adding up the probabilities of the three outcomes *BBG*, *BGB*, and *GBB*.]

$$(\tfrac{8}{12} \times \tfrac{7}{11} \times \tfrac{4}{10}) + (\tfrac{8}{12} \times \tfrac{4}{11} \times \tfrac{7}{10}) + (\tfrac{4}{12} \times \tfrac{8}{11} \times \tfrac{7}{10}) = 3 \times \frac{(8 \times 7 \times 4)}{(12 \times 11 \times 10)} = \frac{28}{55}.$$

Sampling With or Without Replacement

Solving for probabilities in the coin flip and card selection examples above followed a certain pattern, whereas solving for probabilities in the doughnut and student selection examples followed another pattern. When a coin is flipped, the probability of getting a head each time does not depend on how many heads have appeared already; since the cards are replaced, the probability of getting a diamond each time is independent of how many diamonds have been chosen. Such

situations are called **sampling with replacement**, and the selections involved are *independent*. On the other hand, because the doughnuts are not replaced, the probability of getting a jelly doughnut each time depends on how many have already been chosen, and likewise for the probability of getting a boy when students are chosen from a group. Such situations are called **sampling without replacement**, and the selections involved are *dependent*. The differences between probabilities calculated for sampling with replacement as opposed to sampling without replacement may be negligible, or they may be quite large, depending on various background factors.

Question   Suppose there are four cards, including one diamond and three non-diamonds. If two cards in a row are selected *with replacement*, what is the probability that both are diamonds?

Answer   Since the cards are replaced each time, the probability of getting a diamond each time is ¼. Altogether, the probability of getting two diamonds is ¼ x ¼ = 1/16.

◆   ◆   ◆

Question   Suppose there are four cards, including one diamond and three non-diamonds. If two cards in a row are selected *without replacement*, what is the probability that both are diamonds?

Answer   Once a diamond has been selected, there are no more available. It is impossible to get more than one, so the probability of getting two diamonds is zero.

◆   ◆   ◆

Question   If two cards in a row are selected *with replacement* from an ordinary deck that contains 13 diamonds and 39 non-diamonds, what is the probability that both are diamonds?

Answer   Since ¼ of the cards are diamonds, the probability of selecting two diamonds in two selections *with replacement* is ¼ x ¼ = 1/16.

◆   ◆   ◆

Question   If two cards in a row are selected *without replacement* from an ordinary deck that contains 13 diamonds and 39 non-diamonds, what is the probability that both are diamonds?

Answer   In this case, the probability of selecting two diamonds is 13/52 x 12/51 = ¼ x 12/51 = 3/51 = 1/17.

When selecting two cards from a set of 4, the probability of getting two diamonds is 1/16 for sampling with replacement as opposed to 0 for sampling without replacement—quite a dramatic difference. However, when selecting two cards from a set of 52, sampling with or without replacement only involves the difference between 1/16 and 1/17, which is minor. In statistics, relatively small samples are often taken from

a very large group, and the assumption of independence yields accurate probabilities even if the sampling is actually done without replacement. However, always taking independence for granted can give rise to serious errors, so it is important to pay attention to what kind of underlying situation is involved.

Example   We found on page 58 that if three students are chosen at random and without replacement from 12, of whom eight are boys and four are girls, the probability of choosing no boys is $\frac{1}{55}$. Compare this to the probability of choosing no boys if the selection were done with replacement.

Solution   If the selection is done with replacement, since there are four girls in 12 students, the probability of choosing a girl each time is $\frac{4}{12} = \frac{1}{3}$. Choosing no boys means that all three of the students selected must be girls. This event has probability $\frac{1}{3} \times \frac{1}{3} \times \frac{1}{3} = \frac{1}{27} = \frac{2}{54}$, which is more than twice as much as $\frac{1}{55}$.

Experiment: Sampling With or Without Replacement

Materials: 4 slips of paper and a container

On separate slips of paper, write down the names of two boys and two girls; put the four names in a container.

1.   Select, at random, two names in a row *with* replacement. Record whether or not both were girls. Repeat this experiment for a total of 24 times and record the overall proportion of times both names selected were girls.

2.   Select, at random, two names in a row *without* replacement. Record whether or not both were girls. Repeat this experiment for a total of 24 times and record the overall proportion of times both names selected were girls. Does there seem to be a difference in the likelihood of choosing two girls depending on whether the sampling is done with or without replacement?

3.   Calculate the actual probability of selecting two girls' names in both cases, that is, with and without replacement. Compare these probabilities to the proportions occurring in your experiments above.

Expected Value (Optional)

Question   What is the expected number of diamonds selected when three cards in a row are chosen with replacement?

Answer   The probability of choosing a diamond each time is ¼. For three selections we add the probabilities together: ¼ + ¼ + ¼ = ¾. On the average, we expect ¾ diamonds to be selected over the course of three trials.

Alternate Answer   Looking back at our probability distribution on page 53, we see that

$^{27}\!/_{64}$ of the time, 0 diamonds will be picked;
$^{27}\!/_{64}$ of the time, 1 diamond will be picked;
$^{9}\!/_{64}$ of the time, 2 diamonds will be picked; and
$^{1}\!/_{64}$ of the time, 3 diamonds will be picked.

On the average, we expect the number of diamonds picked to be
$(^{27}\!/_{64} \times 0) + (^{27}\!/_{64} \times 1) + (^{9}\!/_{64} \times 2) + (^{1}\!/_{64} \times 3) = {}^{(0 + 27 + 18 + 3)}\!/_{64} = {}^{48}\!/_{64} = \tfrac{3}{4}$

Basic Rule   To find the **expected value** of a random variable, we multiply each possible value by the probability that it will occur and add these up together.

Example   If two doughnuts are chosen at random from a bag containing three filled with jelly and five filled with cream, find the expected number of jelly doughnuts chosen.

Solution   The probability of choosing a jelly doughnut each time varies, depending on how many have already been chosen. But we can still refer back to our probability distribution on page 56 and use our basic rule to find the expected number of jelly doughnuts chosen. Multiplying each possible number of jelly doughnuts chosen by the probability of choosing that many jelly doughnuts and adding them together gives us $^{10}\!/_{28} \times 0 + {}^{15}\!/_{28} \times 1 + {}^{3}\!/_{28} \times 2 = {}^{21}\!/_{28} = \tfrac{3}{4}$ as the expected number of jelly doughnuts chosen.

Note: The expected value is not a probability but a value of a random variable. In this case, the expected value of the random variable "number of jelly doughnuts chosen" could have ranged anywhere from 0 to 2.

Depicting Discrete Probability Distributions: Histograms

Question   What is the probability of getting either zero, one, or two heads when a coin is flipped twice?

Answer   By adding up the probabilities ¼ + ½ + ¼ from the distribution of the random variable for number of heads in two coin flips, we can find the probability of getting either zero, one, or two heads to equal 1.

Alternate Answer   Since zero, one, and two exhaust all possibilities for the number of heads in two coin flips, they take into account the values of this random variable over the full sample space *Fullspace* for the experiment, so together they must have probability equal to 1.

Basic Rule   The sum of all the probabilities in the probability distribution of a discrete random variable must equal 1.

Example   Find the sum of the probabilities in the distribution of the random variable for the number of diamonds picked when three cards in a row are chosen and replaced (see page 54).

Solution $^{27}\!/_{64} + {}^{27}\!/_{64} + {}^{9}\!/_{64} + {}^{1}\!/_{64} = {}^{64}\!/_{64} = 1$.

◆   ◆   ◆

Example   Find the sum of the probabilities in the distribution of the random variable for the number of jelly doughnuts chosen when two doughnuts are selected at random from a bag containing three jelly doughnuts and five cream doughnuts (see page 56).

Solution $^{10}\!/_{28} + {}^{15}\!/_{28} + {}^{3}\!/_{28} = {}^{28}\!/_{28} = 1$.

◆   ◆   ◆

Example   Find the sum of the probabilities in the distribution of the random variable for the number of boys chosen when three students are chosen at random from a group of eight boys and four girls (see page 58).

Solution $^{1}\!/_{55} + {}^{12}\!/_{55} + {}^{28}\!/_{55} + {}^{14}\!/_{55} = {}^{55}\!/_{55} = 1$.

A **probability histogram** is made up of rectangles whose areas represent probabilities. Each rectangle is centered at a possible value of the random variable. If the width of each rectangle is one, then the height must equal the probability that the random variable takes on the value over which the rectangle is centered.

Question How can we depict the probability distributions for the random variables in the previous examples?

Answer The random variable for the number of heads in two coin flips can take on the values zero, one, or two, so we will draw rectangles centered over each of these three numbers. The rectangle over zero should have height ¼, because the probability that the number of heads is zero is ¼. The rectangle over one should have height ½ and the rectangle over two should have height ¼. The histogram looks like this:

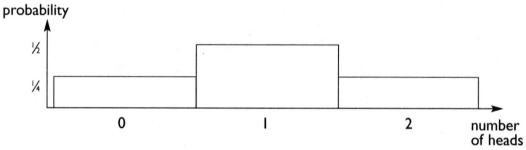

Referring back to the probability distribution for the number of diamonds chosen when three cards are selected and replaced, we can also depict the behavior of this random variable with a histogram:

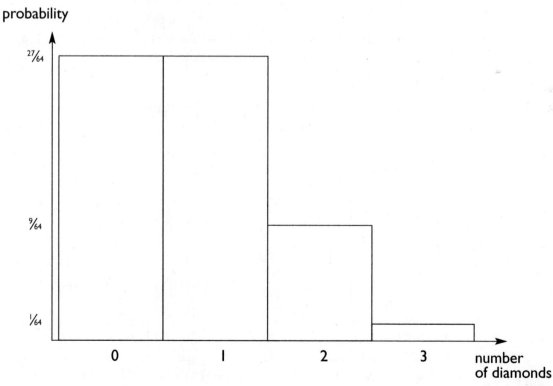

Example   The sample space for the roll of two dice contains 36 possibilities. Counting up the number of ways *n* to get a certain total from 2 through 12, and dividing each time by *N* = 36, we can find the following probability distribution for the random variable "total rolled":

Total Rolled	Probability
2	1/36
3	2/36
4	3/36
5	4/36
6	5/36
7	6/36
8	5/36
9	4/36
10	3/36
11	2/36
12	1/36

Draw a histogram for this distribution.

Solution   Since the random variable can take on any whole number value 2 through 12, we must draw rectangles centered over these numbers. Referring to the given probabilities for heights of the rectangles, we can draw the histogram as follows:

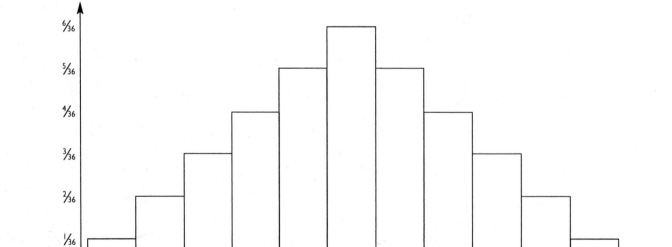

Question   How much area is taken up by all the rectangles in each of the histograms above?

Answer   We first note that the rectangles in our histograms all have width one, so the area of each rectangle just equals its height.

The histogram for number of heads in two coin flips has total area ¼ + ½ + ¼ = 4⁄4 = 1.

The histogram for number of diamonds in three card selections has total area ²⁷⁄₆₄ + ²⁷⁄₆₄ + ⁹⁄₆₄ + ¹⁄₆₄ = ⁶⁴⁄₆₄ = 1.

The histogram for the total rolled with two dice has area ¹⁄₃₆ + ²⁄₃₆ + ³⁄₃₆ + ... + ¹⁄₃₆ = ³⁶⁄₃₆ = 1.

◆   ◆   ◆

Question   A dice game is played where the number of points earned is *double* the number appearing on the roll of a single die. How can a probability histogram be constructed for the number of points earned?

Answer   First we solve for the probability distribution of the random variable "Number of Points Earned":

Number of Points Earned	Probability
2	⅙
4	⅙
6	⅙
8	⅙
10	⅙
12	⅙

We want the *area* of the rectangle over each value to equal the probability that the random variable takes on that value. Since each rectangle will have a width of two, the area will be correct if we make the height of each rectangle equal to half the probability that the random variable takes on each value:

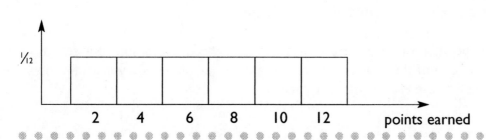

probability
÷ 2

¹⁄₁₂

2   4   6   8   10   12   points earned

Basic Rule	The total area of the rectangles in a probability histogram equals one. This is true because the area of each rectangle gives the probability of the random variable taking on a single possible value, and adding up all the areas is the same as adding up probabilities over the full sample space *Fullspace*.

Example If we drew a probability histogram for the number of boys chosen when three students are selected at random from a group of eight boys and four girls, what would be the total area of the rectangles?

Solution For this, as for any probability histogram, the total area of the rectangles must equal one.

Experiment: Game Show

Materials: 3 envelopes numbered 1, 2, and 3, and a piece of play money for each pair of students [the envelopes should be colored to prevent cheating just as the actual game show would be conducted with doors, not windows!]

Imagine a game show where a fabulous prize has been hidden behind Door #1, Door #2, or Door #3. The host asks the contestant to select a door. After the contestant tells the host which door he or she chooses, the host opens up one of the other doors, showing no prize. Then the host gives the contestant the option of either keeping his or her original selection or switching to the other unopened door. Should the contestant keep the same door or switch; or doesn't it matter?

Simulate this game show by hiding a piece of play money in one of three envelopes and getting a partner to pick an envelope. Show your partner an unpicked envelope which is empty and give him or her the option of switching. Try this several times, then reverse roles and try it a few more times. Does it seem to matter whether you keep the same envelope or switch?

When playing such a game, you want to use a strategy which gives you the highest probability of winning. To find the probability of winning, it is necessary to explore all the possible outcomes in this experiment.

Since the position of the prize can be any one of three doors, and your selection can be any one of three doors, there are, altogether, nine equally likely possibilities. Complete the list on the next page:

(prize door, selected door)	probability
(1,1)	⅑
(1,2)	⅑

Obviously, the host will never show you a door that has the prize. Complete the list of which door the host will open, depending on where the prize is and what door you have selected:

(prize door, selected door)	probability	host shows:
(1,1)	⅑	2 or 3
(1,2)	⅑	3

Since whether or not you win may depend on your strategy, complete a separate chart for each strategy, "KEEP" or "SWITCH":

KEEP

(prize door, selected door)	probability	host shows:	final selection	win or lose?
(1,1)	⅑	2 or 3	1	win
(1,2)	⅑	3	2	lose

SWITCH

(prize door, selected door)	probability	host shows:	final selection	win or lose?
(1,1)	⅑	2 or 3	3 or 2	lose
(1,2)	⅑	3	1	win

What is the overall probability of winning, given that the "KEEP" strategy is used? What is the probability of winning, given that the "SWITCH" strategy is used? Is one strategy better than the other?

Exercise:
Calvin's Test

Suppose Calvin's true-or-false test has four questions, and he answers each totally at random. Then the probability of each question being correct is ½, the same as the probability of being incorrect. Complete the chart which should show all 16 possibilities in the sample space (all combinations of *I*, meaning incorrect, and *C*, meaning correct, for each of the four questions) along with the probability of each outcome and the corresponding number correct. Use the chart to fill out the probability distribution table, which lists possible numbers of correct answers, 0 through 4, and the probability of each of these.

S	Probability	Number Correct
IIII	½ x ½ x ½ x ½ = ¹⁄₁₆	0
IIIC	¹⁄₁₆	1
IICI	¹⁄₁₆	1
ICII	¹⁄₁₆	1
CIII	¹⁄₁₆	1
IICC	¹⁄₁₆	2
CCCC	¹⁄₁₆	4

The probability distribution is:

Number Correct	Probability
0	¹⁄₁₆
1	¹⁄₁₆ + ¹⁄₁₆ + ¹⁄₁₆ + ¹⁄₁₆ = ⁴⁄₁₆
4	¹⁄₁₆

1.   Display this probability distribution with a histogram.

2.   How many correct answers would Calvin need to get *more* than half of the four questions correct?

3.   What is the probability that Calvin gets more than half of the four questions correct, and thus manages to pass the test even though he was only guessing?

4.   (Optional) I personally have serious doubts about the claim that an event has occurred if it has a probability of less than 5% (or 1 in 20 times). If Calvin claims he was only guessing on his true-and-false test, and scored more than two correct out of four, should I have serious doubts about his claim? Explain.

Statistical Inference I (Optional)

Introduction

Statistical inference is the process of gathering information from samples in order to infer something about the larger group from which the samples were taken. For the most part, the goal of statistical inference is to gain information about the value of some unknown continuous random variable. This is done either by finding "confidence intervals" or by doing "hypothesis testing," both of which are covered for continuous random variables in a later chapter. Especially for hypothesis testing, the reasoning involved may seem rather complicated to a beginner. Thus, we pave the way here by introducing the thought process required for making a test of hypothesis on a discrete random variable.

Experiment: "Doubtful" Events

Materials: a coin

If an event is described as being "amazing," about what percent of the time do you think it could occur? Tell what each of the following descriptions means to you by assigning a percent or probability to each. [There is no single "right" answer.] Put an "X" next to all the descriptions which you would put under the heading "doubtful."

impossible	unusual
unbelievable	possible
incredible	not uncommon
amazing	common
very unlikely	everyday
extraordinary	usual
rare	ordinary
questionable	likely
pretty unlikely	probable
surprising	almost certain
unexpected	definite

Flip a coin 10 times and make a note of how many times heads turn up. Which of the words from the list above would you use to describe the probability of the outcome which you just witnessed? If someone flipped his own "lucky penny" 10 times and got eight heads, what word would you use to describe the probability of this event? What word would you use for the probability of getting nine heads? Ten heads? Would you at any point consider the probability to be so low that you must doubt that the coin is really balanced?

Testing a Hypothesis

In general, the probability of one specific outcome occurring may be quite small, especially for a continuous random variable or for a discrete random variable when there are many trials.

For example, the probability that a randomly chosen student is exactly 5½ feet tall, down to the last fraction of an inch, is actually zero, because there are an infinite number of possible heights. On the other hand, the probability of a randomly chosen student having a height, say, less than or equal to 5½ feet tall, may be quite large. [Note that height is a continuous random variable.]

As another example, suppose you have a balanced coin: for each toss, heads and tails are equally likely. If the coin is tossed 400 times, there are so many different possible outcomes that even the most likely one (namely, getting exactly 200 heads) has a probability of only 4%, approximately. But the probability of getting between 190 and 210 heads is close to 70%! [Note that the number of heads tossed is a discrete random variable.]

For this reason, it is often appropriate to consider the probability that a random variable takes on a value *over a certain range*, instead of the probability that it takes on one specific value. If we want to decide whether or not to believe Calvin's claim (in the previous exercise) that he guessed on every true-false question, we should consider the likelihood (or unlikelihood) of his getting three or more questions correct, rather than his getting exactly three questions correct.

Question   Suppose someone claims a coin is balanced, and gets two heads in 10 tosses. Should we believe the claim that the coin is balanced? Use the following probability distribution for the number of heads in 10 coin tosses:

Number of Heads	Probability
0	.001
1	.010
2	.044
3	.117
4	.205
5	.246
6	.205
7	.117
8	.044
9	.010
10	.001

Answer   In order to decide whether or not to believe the claim, we should consider the probability of getting two heads *or fewer* (as opposed to the probability of getting exactly two heads). If it is "doubtful" enough (let's say, less than 5% probability) that 10 tosses of a balanced coin would result in *as few as* two heads, then we may have good reason to doubt the claim that the coin is balanced. If it is not so unlikely (let's say, probability 5% or more), then we will believe the claim that the coin is balanced.

To make a decision, all we need to do is look at the above probability distribution for number of heads in 10 tosses of a balanced coin. [This distribution could be found using the methods of the Probabilities chapter, but obviously requires much more work than the distribution for two coin tosses.] We see that the probability of tossing 2 heads or fewer is the sum of the probabilities of tossing 0, 1, and 2 heads: .001 + .010 + .044 = .055 = 5.5%.

In other words, if the coin is truly balanced, then the probability is .055 or 5.5% that in 10 tosses, by chance alone the number of heads would be 2 or fewer. This is a fairly small probability, but we agreed in advance to accept the claim of the coin being balanced as long as the probability was at least 5%, or .05. Since .055 is greater than .05, we conclude that the coin may be balanced as claimed.

Example   On the next page is the probability distribution for the total rolled with three dice. Suppose Abe is playing a dice game with Benedict, who may be cheating whenever Abe's back is turned. Abe decides to deliberately look away on the next roll and make an outright accusation if the probability is less than 2% that Benedict would happen to roll a total at least as high as it turns out to be. While

Abe pretends to tie his shoe, Benedict rolls the three dice and gets a total of 17. Does Abe have reason to doubt that Benedict is playing fair? [Remember to use 2% as the cut-off probability.]

Total Rolled	Probability
3	$\frac{1}{216}$
4	$\frac{3}{216}$
5	$\frac{6}{216}$
6	$\frac{10}{216}$
7	$\frac{15}{216}$
8	$\frac{21}{216}$
9	$\frac{25}{216}$
10	$\frac{27}{216}$
11	$\frac{27}{216}$
12	$\frac{25}{216}$
13	$\frac{21}{216}$
14	$\frac{15}{216}$
15	$\frac{10}{216}$
16	$\frac{6}{216}$
17	$\frac{3}{216}$
18	$\frac{1}{216}$

Solution

The probability of rolling a total at least as high as 17 with three dice is $\frac{3}{216} + \frac{1}{216} = \frac{4}{216} = .0185$. Since .0185 is less than .02, and Abe doubts a claim which results in an event having probability less than 2%, Abe does have reason to conclude that Benedict is cheating.

◆　　◆　　◆

Example

Still using 2% as the cut-off probability, should Abe accuse Benedict of cheating if he rolls a 16 while Abe is tying his shoe?

Solution

The probability of getting a total at least as high as 16 in the roll of three dice is $\frac{6}{216} + \frac{3}{216} + \frac{1}{216} = \frac{10}{216} = .044$. Since .044 is greater than .02, there is not enough evidence for Abe to accuse Benedict of cheating.

Question

What is the claim to be investigated in our problem of two heads in 10 coin tosses?

Answer

The claim to be investigated is that the coin is balanced.

In statistics, a claim that is to be investigated or tested is called a **hypothesis**. In the same spirit as the law which considers a person to be innocent until proven guilty, we generally begin with a hypothesis that claims normality, "status quo,"

nothing different, or nothing unusual. We will doubt the claim only if it means something is happening which would be very unlikely to happen by chance alone—in other words, if something is "doubtful" as far as the laws of probability are concerned.

Example What is the hypothesis to be tested in the dice game between Abe and Benedict?

Solution The hypothesis to be tested is that Benedict rolls such a high total by chance alone and is not cheating.

◆　　◆　　◆

Example Carmen and Denise are classmates who discover that both enjoy shooting free throws with a basketball. Carmen tells Denise that, overall, she makes 30% of her shots, but Denise suspects that Carmen is just being modest, and actually makes more than 30% of her shots.

1. What is the hypothesis to be tested?

2. Suppose Carmen is successful in nine of her next 15 shots. Below is the probability distribution for number of successes in 15 tries when the probability of success each time is 30%. Use it to decide if Denise has reason to doubt Carmen's claim that she only shoots 30%. [Use .05 as your cut-off probability.]

Number of Successes	Probability
0	.005
1	.030
2	.092
3	.170
4	.219
5	.206
6	.147
7	.081
8	.035
9	.012
10	.003
11	.001
12	.000
13	.000
14	.000
15	.000

Solution

1. The hypothesis to be tested is that overall Carmen shoots with no better than 30% accuracy.

2. According to the given distribution, the probability that someone who shoots each time with only 30% accuracy would be successful in at least nine out of 15 throws is just .012 + .003 + .001 + .000 + .000 + .000 + .000 = .016, or 1.6%. Since .016 is less than .05, Denise has reason to conclude that Carmen is unlikely to shoot 30% as claimed, but actually does better than that.

[Notice that the above situations, besides involving a claim to be tested, also involve some sort of trial whose result must be taken as fact. Thus, we are *not* in a position to doubt that someone flips two heads in 10 coin tosses, but we may doubt that the coin is balanced. We cannot doubt that Carmen makes nine out of 15 free throws, but we may doubt that she shoots with an overall success rate of 30%.]

Beginning Statistics

Introduction

The science of **statistics** concerns itself with the interpretation of lists of number or category values which are called **data**. In order to make sense of a set of data, we need to find appropriate ways to summarize it, using numbers, words, or pictures.

Beginning Experiment: Summarizing Ages

Materials: list of ages (in months) of class members

Glancing at a list of the ages (in months) of class members, write a few sentences to summarize the list informally.

Numerical Summaries of Data: The Median and the Five Number Summary

Question How can we summarize the information in a list of heights of class members?

Answer One way to summarize the data set is to tell where the data are *centered*. For instance, we could report the middle height value.

Basic Rule One measure of center of a numerical data set is the **median**. If there are an odd number of values, the median is the middle value. If there are an even number of values, the median is halfway between the two middle values.

Example Suppose there are 20 people in the class, with heights (in inches) given on the next page:

Person	Height in Inches
Kyle	54
Allen	54
Kerry	54
Theresa	55
Andrew	55
Evelyn	56
Micah	56
Karen	56
Justin	57
Neil	57
Emily	58
Brian	58
Derek	58
Adam	59
Stephen	59
Herman	59
Zachary	60
Nancy	62
Sean	62
Benedikt	65

Find the median of this data set.

Solution The middle of 20 values is halfway between the 10th and 11th: the median height is $\frac{(57 + 58)}{2}$ = 57.5 inches.

◆　　◆　　◆

Example Below is a list of heights of my family members. Find the median height value.

Person	Height in Inches	Person	Height in Inches
Rebecca	29	Nancy	62
Luke	36	Jessie	64
Greta	45	Sue	65
Carolyn	45	Bill	67
Nils	46	Grandpa	68
Natalie	55	B.J.	72
Marina	56	Ken	72
Mattie	56	Frank	72
Andreas	57	Mark	75
Grandma	58		

Solution   Since there are 19 values, the median is the 10th value, or 58 inches.

Question   How do heights of my family members compare to heights of class members?

Answer   Since the median height of class members is 57.5 inches, and the median height of family members is 58 inches, the medians of the two data sets are quite close.

◆   ◆   ◆

Question   Since the medians are close, can we say that the distribution (i.e., the pattern of variation) of heights of my family members is similar to that of class members?

Answer   The medians are close, but family members' heights are much more spread out.

◆   ◆   ◆

Question   How can we use numbers to summarize a data set and include information about how much the data are spread out?

Answer   One way to give information about the spread of a data set is to tell what the smallest and largest values are. Even more information is provided by telling the middle value of the lower half of the data set and the middle value of the upper half.

The middle value of the lower half of a data set is called the **first quartile** because a quarter of the data values fall at or below it. The middle value of the upper half of a data set is called the **third quartile** because three quarters of the data values fall at or below it. [The median is actually the **second quartile**.]

Basic Rule	One good numerical description of a data set is the **five number summary**. This is the list, in order, of the smallest value, the first quartile, the median, the third quartile, and the largest value.

Example   Give the five number summary for heights of class members and for heights of my family members.

Solution   The five number summary values for class and family members are:

Group	Smallest	First Quartile	Median	Third Quartile	Largest
Class	54	55.5	57.5	59	65
Family	29	46	58	68	75

At a glance, we can see that the heights of family members are much more spread out than heights of class members.

Graphical Summaries of Data: Stemplots, Box-and-Whiskers Plots

Question   How can we draw a picture to illustrate the pattern of variation in heights of my family members?

Answer   We can write out all the numbers, organizing them in a way that shows the pattern in their variation. One of the most common ways of doing this is to construct a *stemplot*.

Basic Rule   A **stemplot** contains a vertical list of *stems* which include all but the last digit of each number value. The last digit is the *leaf*. Each stem is followed by a horizontal list of leaves. If the stemplot is not spread out enough to show a clear pattern, stems may be **split** (for example, into two or into five stems). If the stemplot is too spread out to show a clear pattern, the last digit may be **truncated** (cut off).

Example   Construct a stemplot for the heights of my family members.

Solution   Since the range is from 29 to 75 inches, we use the 10s digits 2 through 7 as our stems. Each stem is followed by a leaf for each value that has that stem's 10s digit.

```
2 | 9
3 | 6
4 | 5   5   6
5 | 5   6   6   7   8
6 | 2   4   5   7   8
7 | 2   2   2   5
```

◆   ◆   ◆

Example   Can we use a similar construction for the stemplot of class members' heights?

Solution   Since class members' heights are concentrated in the 50–60 inch range, the stems should be split in order to give a better picture of the pattern of variation. Splitting each stem five ways, we would use a stem for leaves 0 and 1; another for leaves 2 and 3; another for leaves 4 and 5; another for leaves 6 and 7; and another for leaves 8 and 9.

```
5 | 4   4   4   5   5
5 | 6   6   6   7   7
5 | 8   8   8   9   9   9
6 | 0
6 | 2   2
6 | 5
```

Question   Suppose we made a stemplot of ages of everyone in the classroom, including the teacher. What pattern would we see in the stemplot?

Answer   All of the ages would be close to the median age, except for the teacher's age, which would be extremely far from the median.

Basic Rule   An **outlier** is an extreme value in the data set, far from the rest of the data. Outliers can be seen clearly on a properly constructed stemplot.

Example   On a small shelf, I keep some books I have read recently. The number of pages in those 14 books are:

783   174   144   173   240   214   231   463   174   237   368   116   184   564

How can we display and describe the distribution of the number of pages in those books? Are there any outliers?

Solution   First we list the numbers in order from smallest to largest:

116   144   173   174   174   184   214   231   237   240   368   463   564   783

If we constructed a stemplot with each last digit as a leaf, the stems would have to extend from 11 all the way to 78, resulting in a long, scraggly plot with no clear pattern to be seen. Instead, we truncate (cut off) the last digit and work with the values

11   14   17   17   17   18   21   23   23   24   36   46   56   78

Now we can construct a stemplot with stems from 1 to 7:
```
1 | 1   4   7   7   7   8
2 | 1   3   3   4
3 | 6
4 | 6
5 | 6
6 |
7 | 8
```

The median is between the seventh and eighth values: the truncated values are 21 and 23, corresponding to the original values 214 and 231. The median is $^{(214 + 231)}\!/_2 =$ 222.5. The smaller values are all quite close to the median, but the larger values are spread further away. In particular, the value 783 may be considered to be an outlier because it is especially far from the rest of the values.

To summarize, we can say that half of the books I read were no longer than 222.5 pages. The shorter books weren't much shorter than 222.5 pages, but the longer books had considerably more pages, especially the one with 783 pages.

Question   How can we use stemplots to compare heights of male and female class members, as given?

Females: 54   55   56   56   58   62
Males: 54   54   55   56   57   57   58   58   59   59   59   60   62   65

Answer   Constructing the stemplots back-to-back will make it easier to compare them:

Females			Males		
5	4	5	4	4	5
6	6	5	6	7	7
	8	5	8	8	9 9 9
		6	0		
	2	6	2		
		6	5		

Basic Rule   A **back-to-back stemplot** may be used to compare two data sets. Both plots share the same vertical list of stems. The leaves for one data set precede the stems right to left, and the leaves for the other data set follow the stems left to right.

Example   Make a back-to-back stemplot to compare heights of male and female members of my family, as given:

Females: 29   45   45   55   56   58   62   64   65
Males: 36   46   56   57   66   67   68   72   72   72   75

Solution   The males and females share the stems 2 through 7. We can list the leaves for female members' heights starting at the left of the stems, and for male members' heights starting at the right of the stems:

Females				Males			
	9	2					
		3	6				
5	5	4	6				
8 6	5	5	6	7			
5 4	2	6	6	7	8		
		7	2	2	2	5	

Males tend to be somewhat taller than females. Both data sets have most values pretty close to the median, which is 55 inches for the females and 67 inches for the males. Neither data set is noticeably more spread out than the other.

Question   Referring back to our stemplot on page 81 for the numbers of pages in a set of books, does the plot appear to be fairly well-balanced about the middle, or is it lopsided?

Answer   The larger values are spread farther away from the middle than the smaller values, resulting in a lopsided shape.

Basic Rule   The **shape** of a distribution can be described as follows: if numbers in a data set are fairly well-balanced on either side of the median, then we say they are distributed in a **symmetric** way. If not, we say they are **skewed**. In particular, if the smaller values are spread further away from the middle, we say the data set is skewed toward smaller values (or "skewed left," where we are thinking of the stemplot as being tilted, with small values to the left and large values to the right). If the larger values are spread farther away from the middle, we say the data set is skewed toward larger values (or "skewed right").

Example   The stemplot below shows students' scores on their first test in a ninth grade French class. Describe the data.

5	4	4										
5	7											
6	1	3										
6	8											
7	1	3										
7	5	6	7	8								
8	1	1	1	1	4							
8	5	5	6	6	7	7	7	7	8	8	8	8
9	1	3	4									
9	6											
10	0	0										

Chances Are ...

Solution   Since there are 35 score values, the middle (median) of the data set is at the 18th value, which is 85. There are no outliers, but the smaller scores are noticeably spread out further away from the median: the data set is skewed left, toward smaller values.

◆   ◆   ◆

Example   Now suppose the students from our example above are separated into two groups, depending on whether or not they had studied French before in middle school. Describe the data.

```
Did Not Study Before        Did Study Before
              4   4 | 5 |
                  7 | 5 |
              3   1 | 6 |
                  8 | 6 |
                  1 | 7 | 3
              8   7 | 7 | 5   6
                  1 | 8 | 1   1   1   4
          8   8   6 | 8 | 5   5   6   7   7   7   7   8   8
                  3 | 9 | 1   4
                    | 9 | 6
                  0 | 10| 0
```

Solution   There are 15 students who did not study French before. Their median score is the eighth one, or 77. The data set is spread out quite a bit on both sides of the median, without any noticeable skewness or outliers: it is a fairly symmetric data set.

There are 20 students who did study French before. Their median score is halfway between the 10th and 11th scores: $(86 + 87)/2 = 86.5$, quite a bit higher than the median for the other students. Their lowest score, 73, is much higher than the lowest score for the other students, which was 54. However, the highest score, 100, was the same for both groups. The scores for students who had studied French before are not nearly as spread out as those of the students who hadn't. The shape of their stemplot is also quite symmetric.

The skewness that appeared in the first stemplot came about because the new students, as a group, tended to have lower scores. Examining the two groups separately gives us a much more detailed picture of the students' performances.

Question   What if a data set is so large that a stemplot is impractical? What if we want to compare more than two data sets?

Answer   Instead of displaying all the values in a data set, we could simply display a few key values—for instance, the values of the five number summary.

Basic Rule   A **box-and-whiskers plot** is a graphical display of the five number summary.

1.   The bottom "whisker" extends down to the smallest value.
2.   The bottom of the box starts at the first quartile.
3.   A line is drawn through the box at the median.
4.   The top of the box is at the third quartile.
5.   The top whisker extends up to the largest value.

By this construction, the box shows where the middle half of the data values are. Several boxplots may be drawn side-by-side to compare two or more data sets.

Five Number Summaries for Pulse Rates

Group	Smallest	First Quartile	Median	Third Quartile	Largest
Newborns	120	140	148	156	160
Aged 60-65	60	68	72	78	106
Class Members					

Example   1.   Draw side-by-side boxplots to compare pulse rates of newborns, people aged 60–65, and class members (to be filled in), using the five number summaries given above.

2.   Suppose a doctor is told by his assistant that a new patient has a pulse rate of 113. Should he be alarmed?

Solution   1.   Boxplots for the pulse rates of the group of newborns and the group of people aged 60–65 are shown below. A boxplot for class members may be added, and all three groups may be compared.

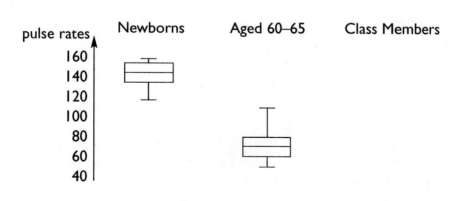

Pulse rates of newborns are clearly higher than those of people aged 60–65. In fact, *the* slowest of the newborn pulse rates (namely, 120) is still faster than *the* fastest pulse rate for people aged 60–65 (namely, 106). The long bottom whisker on the newborn plot indicates that newborn pulse rates have small outliers or are skewed toward smaller values. The long top whisker on the aged 60–65 plot indicates that their pulse rates have large outliers or are skewed toward larger values. Neither distribution is noticeably more spread out than the other.

2. If the patient is an infant, the doctor could well be concerned that the pulse rate is too low; if the patient is elderly, then the pulse rate is unusually high. If the patient is an older child or a young adult, then a pulse rate of 113 may be considered normal.

Finishing Experiment:
Summarizing Ages

1. Use what you have learned so far in this chapter to examine the distribution of class members' ages (in months) in an organized fashion.
   - Display the ages in a stemplot.
   - Circle the values that make up the five number summary: the youngest, the middle of the younger half, the middle, the middle of the older half, and the oldest.
   - Display the five number summary with a boxplot.
   - Write a few sentences to summarize the distribution of ages. Be sure to include mention of center, spread, and shape.

2. Compare this formal summary to the informal one you wrote earlier. Which do you prefer?

Exercise:
Comparing Distributions

Collect your own set of data—between 10 and 20 values would be a good size to work with. Then collect another set that could be compared to the first set.

1. Make a back-to-back stemplot for the two data sets.
2. Circle the five number summary values in each stemplot. [As an option, you may display these with side-by-side boxplots.]

3. Summarize and compare the distributions as thoroughly as possible, including mention of center, spread, and shape.

If you want, you can use one of these ideas for two data sets:

*   Select some of your own books and make a list of how many pages each has; ask a family member if you can make a list of how many pages are in a selection of *his or her* books and compare with a back-to-back stemplot; or
*   At the grocery store, make a list of calorie contents per serving of some kids' cereals. Then make a list of calorie contents per serving of some "adult" cereals. Compare them with a back-to back stemplot; or
*   Get a bunch of pennies (at least 20) and make a list of their dates. Get a bunch of nickels (or dimes, or quarters) and make a list of their dates. Compare the two data sets and see if pennies tend to be any older or newer than other coins.
*   The newspaper's sports page is full of data. For instance, you can compare batting averages of your favorite baseball team members to those of another team. Since stemplots are more easily used to look at distributions of whole numbers, work with the *number* of hits per thousand instead of with decimal values.

Remember, if you are working with numbers in the hundreds, you may want to truncate (leave off) the last digit of each value and only work with the 10s digits. [See the "pages" example on page 81.] If you are working with many 2-digit numbers that have the same 10s digit, you may want to split stems. [See the "ages" example on pages 80–81.]

Measuring Center of Categorical Data: The Mode

Question   How can we briefly summarize the zip codes of class members?

Answer   The numbers on a zip code correspond to different parts of the city. The order of the numbers is not very meaningful, so reporting the median value wouldn't make sense. The best we can do is to tell which zip code is used by the most class members.

> **Basic Rule**   The **mode** of a data set is the value that occurs most frequently. It is often used to summarize a categorical data set.

Example   Briefly summarize this data set on different ways Americans celebrate their dogs' birthdays:

Birthday Celebration	Number Practicing
Give dog a bone	964,708
Give dog a cake	1,850,665
Give dog ice cream	1,102,524
Give dog a party with other dogs or pets	659,545
Give dog a toy	1,801,445
Give dog a special treat	4,567,598
Make dog a special meal	1,978,636
Sing or wish dog happy birthday	698,921
Take dog to a favorite place	393,758
Take photographs	216,567

("The Unofficial US Census" by Tom Heyman.)

Solution The mode of this data set, or most common celebration, is "Give dog a special treat."

Another Measure of Center: The Mean

Question Three children agreed before a chocolate egg hunt to divide up evenly all the eggs they would find. If Nils found 2, Marina found 4, and Andreas found 12, how can the eggs be divided evenly?

Answer The three children can redistribute the eggs as follows: Andreas gives 2 of his eggs to Marina and 4 to Nils. This balances things out so that everyone has 6 eggs.

Alternate Answer The total number of eggs, $2 + 4 + 12 = 18$, could be divided evenly among the three children so that each gets $\frac{18}{3} = 6$ eggs.

◆   ◆   ◆

Question If uniform blocks of equal weight are placed on a see-saw in the positions shown, where should the see-saw be "centered" in order to balance?

2   4   6   8   10   12

Answer The "balance point" of the numbers 2, 4, and 12 is their arithmetic average, or *mean*: $\frac{(2 + 4 + 12)}{3} = 6$. The see-saw will balance around the point 6. [Technically, the see-saw should be weightless, and the weight of each block should be concentrated over a single point. But our construction will give good results as long as the see-saw is light (e.g. a thin ruler) and the blocks are not too wide (say, no more than an inch).]

Basic Rule	The **mean** of a set of values is really the balance point of those values. We calculate it by adding them all together and dividing by the number of values in the set.

Example Suppose another block of the same size is placed on top of the block at position 12. Now how can the see-saw be balanced?

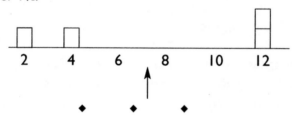

Solution The balance point is the mean of the numbers 2, 4, 12, and 12, namely $\frac{2 + 4 + 12 + 12}{4} = 7\frac{1}{2}$. The see-saw will balance if we "center" it over $7\frac{1}{2}$:

◆   ◆   ◆

Example What is the mean height of the 19 members of my family?

Solution The mean height of my family members is

$$\frac{(29 + 36 + 45 + 45 + 46 + 55 + 56 + 56 + 57 + 58 + 62 + 64 + 65 + 67 + 68 + 72 + 72 + 72 + 75)}{19} = 58 \text{ inches}$$

◆   ◆   ◆

Example 1. How would the mean age of all class members (*excluding* their teacher) compare to their median age?
2. How would the mean age of all class members (*including* their teacher) compare to their median age?

Solution 1. We can expect the distribution of class members' ages to be fairly symmetric without any major outliers, and the mean age should be close to the median.
2. If the teacher's age is included, it will clearly be a high outlier. Averaging it in with the rest of the ages will produce a large value for the mean. The median, on the other hand, will only shift up slightly. Therefore, the mean in this case will be considerably higher than the median.

Choosing a Measure of Center

Question   A class of students was asked which Lego™ set in a catalog looked the best, identifying each by number. What measure of center is most appropriate to summarize the data?

Answer   The mode is most appropriate: we should tell the number of the Lego™ set preferred by the most students.

♦   ♦   ♦

Question   If our data set consists of the ages (in months) of class members, what would be the best measure of center to use in order to summarize the data?

Answer   The mean, or arithmetic average, would probably be the best measure of center, because it contains information from each and every class member.

♦   ♦   ♦

Question   If our data set consists of the ages of everyone in the classroom, including the teacher, what would be the best measure of center for describing the data?

Answer   Averaging in the teacher's age would "pull up" the value of the mean, giving the false impression of the students being older than they actually are, just because of one extremely high age value. The median would be a better measure of center in this case.

Basic Rule   The mode is the best measure of center for counts of categorical data, or for data in the form of numbers for which arithmetic wouldn't make sense.

The mean, or arithmetic average, is usually the best measure of center for numerical data because it contains the most information.

If a data set is skewed, or if one or more outliers would have a strong influence on the value of the mean, then the median would probably be a better measure of center.

Example   For each of the following data sets, decide which measure of center would be best.
1.   The beginning letter of each class member's first name;
2.   The number of letters in each class member's first name;
3.   The number of hours per week that each student in a certain class watches television;

4. The amount paid for each of the homes that were bought in Pittsburgh during a particular month.

Solution
1. The mode is the best measure of center because this is a categorical data set.
2. The mean is the best measure of center because this is a numerical data set which wouldn't necessarily have extreme outliers or skewness.
3. The mean is the best measure of center because this is a numerical data set which wouldn't necessarily have extreme outliers or skewness.
4. The median is the best measure of center because this data set is likely to be skewed toward higher home prices or have high outliers.

Experiment:
Mean Equals Balance Point

Materials: light rulers, small heavy blocks (e.g. Duplo™ blocks stacked in units of three together) or sticks of butter or margarine, and a small block or die for the base

Calculate the mean for each of the following sets of numbers. Then use blocks on a light ruler to verify that the mean is the balance point. [One thin block underneath can serve as the "fulcrum," or balance point. If appropriate blocks are not available, sticks of butter or margarine can be balanced on a ruler with a small die as the fulcrum. Because of the added weight of the ruler, this experiment works best when the blocks are spread out on both sides of the ruler.]

1. 1, 11
2. 1, 6, 11
3. 1, 3, 11
4. 1, 9, 11
5. ½, 11½
6. ½, 6, 11½
7. ½, 3, 11½
8. ½, 3½, 11

9. ½, 2, 11
10. 1, 7, 11½
11. 1, 4, 11
12. 2, 6, 10
13. 2, 9, 10
14. ½, 2½, 11
15. 1, 10, 10

Exercise:
Calculating Means

Calculate the mean values for each of your two data sets from the previous Exercise on page 86. Compare them to the median values. Which would be a better measure of center in each case? [Use your stemplots and boxplots to decide.]

Alternative: Calculate the mean number of pages in 14 books with the following numbers of pages, and compare it to the median. Which would be a better measure of center? [First make a stemplot of the data.]

116
144
173
174
174
184
214
231
237
240
368
463
564
783

Sample Mean and Sample Proportion

Introduction

Besides its meaning as a science, the word "statistics" has another meaning. A **statistic** can also be something we measure from data, like the mean height of a group of class members or the proportion of class members who are female. In contrast to statistics, we will also discuss **parameters**, which are more abstract. For example, the mean height of all children in your age bracket or the proportion of all children who are female would be parameters. It is often important to know—at least approximately—the values of such parameters, although it would not be practical to try to measure them for a very large group.

In statistics (the science) we use statistics (the measurements) to estimate parameters. In other words, we use information from a sample to draw conclusions about the larger group from which the sample was obtained. This larger group is called the **population**. The population may be of unknown size, or even infinite, but it must be fairly represented by the sample in order for our estimates to be reliable.

Examining Statistics and Parameters

Question   If you pick three cards from an ordinary deck of 52, shuffling and replacing each time before picking the next card, what is the mean number value on those three cards? [We will count ace as 1, jack as 11, queen as 12, king as 13, and the rest as numbered.]

Answer   I personally picked a two, a queen, and a three, averaging $(2 + 12 + 3)/3 = 17/3 = 5\frac{2}{3}$.

If you try this yourself, you'll almost certainly get a different mean. In any case, we can call this calculated average the *sample mean* number value. It is a *statistic* because we actually measure it from the data. [My data were the numbers 2, 12, and 3. What were yours?]

Basic Rule	If we have a sample of data values, we find the **sample mean** value, a statistic, by adding them all together and dividing by the number of values.

Example — The heights of boys and girls in a particular class can be thought of as samples taken from a larger group of boys and girls (for example, from all the boys and girls who have ever taken the class). What is the sample mean height of boys in your statistics class? What is the sample mean height of girls in your statistics class?

Solution — We find the sample mean height of boys by adding up all their heights and dividing by the number of boys. The same can be done for the girls.

Question — If you could go on and on forever picking (and replacing) cards in a deck, what would be the average number value on those cards?

Answer — Because a deck contains four each of the values 1 through 13, we know that ultimately we should average a value of $\frac{4 \times (1 + 2 + \ldots + 13)}{52} =$ 7. But 7 is not a statistic because we cannot really pick cards forever and measure their mean. In this case, 7 is a *parameter* because it describes the mean value of the whole population of infinite card selections.

Basic Rule	The arithmetic average of all the values in an entire population is simply called the **mean** of that population. It is a parameter which we may not be able to actually measure.

Example — What is the mean height of all 11-year-old boys in the United States?

Solution — A health reference book reports that 11-year-old boys have a mean height of 56.4 inches. This is a parameter, not a statistic, because it describes the entire population of 11-year-old boys, not all of whom have actually been measured.

Question — If you pick four cards from an ordinary deck, replacing and shuffling each time before picking the next card, what is the proportion of diamonds selected?

Answer — I personally picked four cards and got two diamonds, so my *sample proportion* of diamonds was ¼ = ½. The sample proportion could very well have a different value when you yourself measure it.

Basic Rule	If we can classify each observation in a sample as belonging or not belonging to a certain category, then the **sample proportion** of observations in that category is the number of observations in the category divided by the total number in the sample.

Example	If we consider a particular class to be a sample taken from some larger population, what is the sample proportion of girls in the class?

Solution	We divide the number of girls by the total number of students in the class to get the sample proportion of girls.

Question	If we went on and on forever picking and replacing cards in an ordinary deck, what would be the proportion of diamonds picked?

Answer	Since ¼ of the cards in an ordinary deck are diamonds, we know that ultimately the proportion of diamonds picked should be ¼. Here ¼ is a parameter, not a statistic, because we have not actually measured it. It describes the whole population of infinite card selections.

Basic Rule	The parameter which describes the fraction of all members in a population which fall into a certain category is simply called the **proportion**.

Example	What proportion of all 11-year-olds are girls?

Solution	We know the value of this parameter to be approximately ½, although we have not actually measured it.

Sample Mean and Sample Proportion as Estimates

Question	How can we estimate the mean height of all 11-year-old girls?

Answer	If we have no reason to believe that the girls in a particular class of 11-year-olds are especially tall or short, it is not unreasonable to use their sample mean height to estimate the mean height of all 11-year-old girls.

Basic Rule	The statistic *sample mean* is often used to estimate the parameter *population mean*.

Example   How could I estimate the weight of a typical 11-year-old boy?

Solution   I could use the sample mean weight of all boys in a particular class of 11-year-olds.

Question   Suppose my friend is teaching a summer sport clinic for 11-year-old female basketball players. Should she use their sample mean height to estimate the mean height of all 11-year-old girls?

Answer   No, because she would almost surely make an overestimate.

The systematic tendency to either over- or underestimate a variable is called **bias**.

Question   If I wanted a good estimate for the average IQ score of all 11-year-old children in Pittsburgh, what should I use?

Answer   Averaging IQ scores of a *random* sample of 11-year-old Pittsburgh children would be the best way to estimate the mean IQ of all 11-year-old Pittsburgh children.

Basic Rule   In a **random** sample, chance alone determines the selection. If we want to use a sample mean to make an estimate for the population mean value of a variable, taking a *random* sample of values of the variable is the best way to avoid bias in either direction.

Example   A gym teacher with 100 fifth grade students needs to estimate the average time required for his fifth graders to run a mile. If he is only able to actually time 10 students, which would be the best method to use?

1.   Pick out 10 fifth graders who seem to have typical running ability and time them running a mile. Their sample mean time should be used as the estimate; or

2.   Assign each student a number from 1 to 100 and then randomly select 10 numbers. The students who were assigned the 10 randomly selected numbers should run and their sample mean time should be used as the estimate.

Solution   Method 2 should be used because a *random* sample provides the best estimate. Otherwise the teacher's selection may result in an estimate that is biased toward faster or slower times.

Question   If I wanted a good estimate for the proportion of words with mistakes that a certain secretary types, which of the following would be a better estimate?

    1.   Ask the secretary to hand in a typed page and find the proportion of mistakes on that page; or

    2.   Pick a page at random from those typed by the secretary and find the proportion of mistakes on that page.

Answer   Method 2 would be better because the sample is random. [In Method 1, the secretary might purposely hand in a page with few mistakes, resulting in bias—in this case, sample proportion would systematically underestimate population proportion.]

Basic Rule   The proportion of an entire population which falls into a certain category may be estimated by taking a *random* sample from the population. The sample proportion of observations which fall into the category can be used to estimate the overall population proportion in that category.

Example   How could we estimate the proportion of red M&Ms™ in a large bag of mixed-color M&Ms™?

Solution   We could close our eyes and take out a random sample handful of M&Ms. The sample proportion of red M&Ms™ could be used as an estimate for the proportion of red M&Ms™ in the whole bag.

Experiment:
Random Numbers

Materials: small slips of paper

1.   Each student thinks of a number "at random" between 1 and 20 and writes it down secretly on a slip of paper. Then the slips are collected and the selected numbers are listed. Do the numbers truly appear to be a random sample, or does there seem to be bias in favor of certain numbers? How can someone pick a number between 1 and 20 which is truly random?

2.   Each student picks three states at random from the following, exactly as they are shown here:

Alabama	Hawaii	Massachusetts
Alaska	Idaho	Michigan
Arizona	Illinois	Minnesota
Arkansas	Indiana	Mississippi
California	Iowa	Missouri
Colorado	Kansas	Montana
Connecticut	Kentucky	Nebraska
Delaware	Louisiana	Nevada
Florida	Maine	New Hampshire
Georgia	Maryland	New Jersey

After each student writes down his or her "random" selection, collect them from all students and count the proportion of times a student selected exactly one state from each of the three columns. Now solve for the probability that a random selection of three states contains exactly one from each column:

•   Find N, the total number of ways to choose 3 states from 30.
•   Find n, the number of ways to choose 1 from 10 states in the first column, 1 from 10 in the second column, and 1 from 10 in the third column.
•   Calculate n/N, the probability that a random selection of three states would contain exactly one from each column.

Do the students' selections of three states appear to be truly random samples? If not, what kind of bias is present?

(Optional) Find the probability that a random selection would result in three states all from the same column (first, second, or third). Did any student "randomly" select all three from the same column?

[These experiments should be performed on as many students as possible in order to determine whether or not there is some pattern in their choice of numbers or states.]

Experiment: Globe Toss

Materials: inflatable globe ball, atlas or geography book

Use a globe ball for the following exercise to estimate the proportion of the Earth's surface that is covered by water.

Toss the ball so that the right index finger of the person who catches it touches a random spot. The sample proportion of times that the finger touches water can be calculated after a reasonable number of tosses (say 20 or 40), and this can be used to estimate the proportion of the entire Earth that is covered by water. Refer to a geography book or atlas to determine how close your estimate is to the actual proportion.

If you used a larger ball, would your estimate tend to be any more or less accurate? Or does the size of the ball not matter?

Relative Frequency Histograms and the Normal Distribution

Introduction

Statisticians observe all sorts of patterns in data, but one pattern is by far the most widely seen: it is called the "normal distribution."

In the Beginning Statistics chapter we began with the concrete idea of a statistic, which we can actually measure, and related it to a parameter, which is more abstract. For instance, the mean number appearing when a die is rolled 10 times is a statistic, but the mean number appearing if a die could be rolled an infinite number of times is a parameter.

In this chapter, we will begin with the concrete *pattern* for the behavior of a variable that has been measured a certain number of times, and relate it to the abstract pattern that we would see if the variable could be measured over and over an infinite number of times. For instance, we will look at the pattern in the total rolled if two dice are tossed 36 times, and then consider the pattern arising if we could toss the dice forever. As another example, we can examine the pattern in heights of a certain number of 11-year-old boys and then think about the pattern we would see in heights of all 11-year-old boys. This last example will show us a typical normal distribution.

Relative Frequency Histograms

We have already used probability histograms to show the pattern for behavior of certain random variables in the long run. If we are interested in the pattern for the behavior of some variable that we actually measure, we may use a stemplot, a boxplot, or a *relative frequency histogram*.

Question Suppose two dice have been rolled 36 times with the following observations for total rolled each time:

2, 3, 3, 3, 4, 5, 5, 5, 6, 6, 6, 6, 6, 6, 7, 7, 7, 7, 7, 7,
8, 8, 8, 8, 8, 9, 9, 9, 9, 9, 10, 10, 10, 10, 11, 11

[The data have been arranged in increasing order to make them easier to work with.] How can we represent the pattern of behavior of the observed totals rolled?

Answer   First we make a chart of each possible total (2 through 12) its frequency, and its relative frequency (in this case, the frequency divided by 36). Then we draw a horizontal line marked with the possible values 2 through 12, and over each value we draw a rectangle whose height equals the relative frequency at which we have observed that particular value.

Total	Frequency	Relative Frequency
2	1	1/36
3	3	3/36
4	1	1/36
5	3	3/36
6	6	6/36
7	6	6/36
8	5	5/36
9	5	5/36
10	4	4/36
11	2	2/36
12	0	0/36

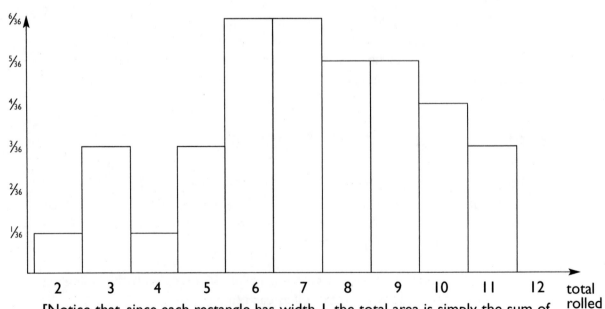

relative
frequency

[Notice that, since each rectangle has width 1, the total area is simply the sum of all the relative frequencies, so it must equal one.]

Question 1. In those 36 rolls of two dice, what was the relative frequency of rolling
- a number less than 4?
- a number between (and including) 5 and 8?
- a number greater than or equal to 10?

2. Find the following values for total rolled:
- The lowest ¼ of the rolls were at or below which total value?
- The middle ⅔ of the rolls were between which total values?
- The highest 50% (or ½) of the totals were higher than which value?

Answer 1. By summing up areas of rectangles over the appropriate values, we find the answers to the first questions:
- The relative frequency of rolling a number less than 4 was ³⁄₃₆.
- The relative frequency of rolling a number between 5 and 8 was ²⁰⁄₃₆.
- The relative frequency of rolling a number greater than or equal to 10 was ⁶⁄₃₆.

2. By finding the values for which the corresponding rectangles have a given total area, we can answer the next questions:
- The lowest ¼ of the rolls (that is, the lowest 9 of the 36 rolls) were at or below the total value 6.
- The middle ⅔ of the rolls (that is, the middle 24 of the 36 rolls) were in between the values 5 and 9.
- The highest ½ of the rolls (that is, the highest 18 of the 36 rolls) were higher than 7.

Basic Rule If a relative frequency histogram for a variable is constructed accurately and the total area of its rectangles equals one, then the relative frequency with which the variable takes on any set of values equals the total area of the rectangles that are centered over those values.

Example The relative frequency histogram on the next page has been constructed using the heights of boys in a particular statistics class.

1. What is the relative frequency that a boy in this class is
- Shorter than 55 inches?
- Between 57 and 59 inches?
- At least as tall as 60 inches?

2. Find the following height values:
   - The shortest ½ of the boys in this class are at or below which height?
   - The middle ¹⁰⁄₁₄ are between which height values?
   - The tallest ½ are taller than which height value?

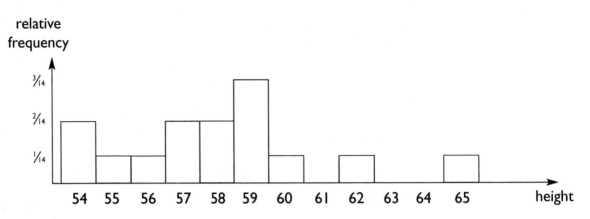

Solution

1. The first questions can be answered by summing up the areas of the rectangles over the appropriate values.
   - The relative frequency is ²⁄₁₄ = ⅐ that a boy in this class is shorter than 55 inches.
   - The relative frequency is ⁷⁄₁₄ = ½ that a boy in this class is between 57 and 59 inches.
   - The relative frequency is ³⁄₁₄ that a boy in this class is at least 60 inches tall.

2. The next questions can be answered by finding the values for which the corresponding rectangles have a given total area.
   - The shortest ½ (or ⁷⁄₁₄) of the boys are at or below 58 inches.
   - The middle ¹⁰⁄₁₄, all but the shortest two and the tallest two, are between 55 and 60 inches.
   - The tallest ½ (or ³⁄₁₄) are taller than 60 inches.

The Normal Distribution

There is an important difference between the variable for the total rolled on two dice and the variable for heights of boys. Because we can count out all the possibilities for value of total rolled (2, 3, 4, ..., 11, and 12), we can say that the total rolled is a *discrete* variable. Heights of boys, on the other hand, don't just take on whole-inch values; they may take on any of all the infinite values in between, continuing from each whole-inch value to the next. Therefore, height is a *continuous* variable.

If we merely take a (finite) sample of values for each of these variables, the relative frequency histograms are constructed similarly. On the other hand, if we imagine the behavior of these variables for a population of infinite size and if heights could be measured to the highest level of accuracy, the behavior of the continuous variable for height should be represented by a smooth curve over the continuous line of values, not by rectangles centered over particular values.

Question   If we continued on and on forever rolling two dice, what pattern would the relative frequency histogram show?

Answer   Because we know the probability of each total in the long run, we can say that the relative frequency histogram ultimately should resemble the probability histogram which we drew on page 64. The area of the rectangles over any interval of numbers shows the probability of rolling any of the numbers in that interval:

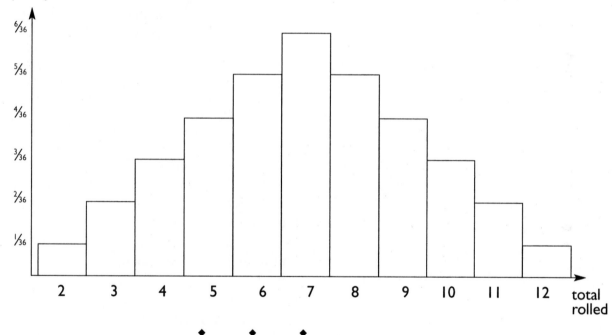

Question   If we were able to measure heights of all 11-year-old boys, not just those in a particular class, what pattern would we see for the variable "height"?

Answer   Scientists discovered such patterns hundreds of years ago. If the variable "height" were measured for an unlimited number of 11-year-old boys, the relative frequency histogram would ultimately resemble the *normal* curve shown on the next page. Now the area under the curve over any interval of height values shows the probability that a randomly selected 11-year-old boy would have a height in that interval of values.

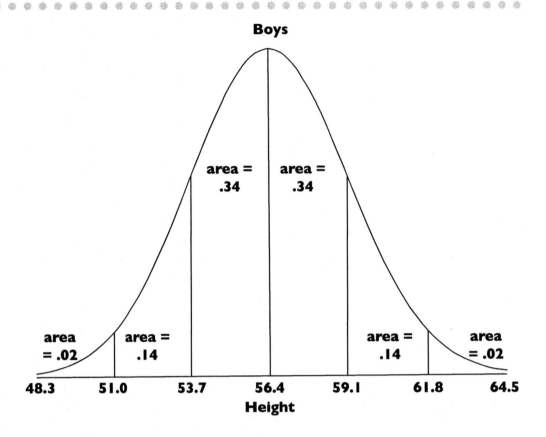

Boys

area =
.34

area =
.34

area
= .02

area =
.14

area =
.14

area
= .02

48.3   51.0   53.7   56.4   59.1   61.8   64.5

Height

Basic Rule	The long-run pattern of behavior of many variables (such as heights, weights, measurement errors, and test scores) can be represented by a **normal** curve, and we say these variables have a **normal distribution**. The normal curve is symmetric—that is, it has mirror images on either side of the middle. It is bell-shaped, showing that there is a high probability that such a variable takes on values close to its mean, and a low probability that it takes on values far from its mean. [Because the normal distribution is symmetric, the mean is equal to the median: the average or expected value occurring actually equals the middle value.]
	The area under the normal curve for a variable between any two values gives the probability that the variable falls somewhere between those values. The total area under the normal curve must equal one.

Example   1.   What is the probability that a randomly chosen 11-year-old boy is
- Shorter than 51.0 inches?
- Between 53.7 and 59.1 inches tall?
- Taller than 59.1 inches?

2.   Find the following height values:
  * 50% of all 11-year-old boys are shorter than what height?
  * The middle 96% of 11-year-old boys have heights between what values?
  * Practically none of the 11-year-old boys are taller than what height?

Solution   1.   By solving for areas under the normal curve, we find:
  * The probability that a randomly chosen 11-year-old boy is shorter than 51.0 inches is .02.
  * The probability that a randomly chosen 11-year-old boy has a height between 53.7 and 59.1 inches is .34 + .34 = .68.
  * The probability that a randomly chosen 11-year-old boy is taller than 59.1 inches is .14 + .02 = .16.

2.   By finding height values which correspond to the given relative frequencies, we see that:
  * 50% of all 11-year-old boys are shorter than 56.4 inches because the area to the left of 56.4 is .02 + .14 + .34 = .50.
  * The middle 96% of 11-year-old boys have height between 51.0 and 61.8 inches because .96 = .14 + .34 + .34 + .14 is the area between 51.0 and 61.8.
  * Practically none of the 11-year-old boys are taller than 64.5 inches.

◆   ◆   ◆

Example   Show that the total area under the normal curve for height of 11-year-old boys equals one.

Solution   The total area is .02 + .14 + .34 + .34 + .14 + .02 = 1.00.

Experiment:
Summarizing Grades

Grade	Points
A	4.00
A-	3.75
B+	3.25
B	3.00
B-	2.75
C+	2.25
C	2.00
C-	1.75
D+	1.25
D	1.00
D-	0.75
F	0.00

Four first-year college students have grades reported as follows:

Jessica: A, B, B, B, B, B, B, C
Kevin: A, A, A, A, C, C, C, C
Leah: A, A, A, B, B, B, B, B, B, B, B, F
Marvin: A, B+, B+, B+, B+, B-, B-, B-, B-, C

1. Write a few sentences that informally describe the performances of those four students.

2. Now use what you have learned in the previous chapters to give a statistical summary of the four students' performances:
    *   List the numerical point assignments for each set of grades.
    *   Report the mean and median for each set of grades.
    *   Draw four histograms, one for each student's grades.
    *   Describe the spreads and shapes of the four distributions as displayed by your histograms.
    *   In what ways are the students' performances alike? In what ways are they different?
    *   Which student's (or students') grades have a distribution which seems closest to "normal"?

3. (Optional) Do a statistical summary of your most recent report card grades. Compare them to past report card grades.

Exercise:
Histogram and Normal Curve for Heights

1. Use a class survey to construct a relative frequency histogram for heights of boys or girls in your class.

2. Use the relative frequency histogram for heights of boys or girls in your class to find the relative frequency that a boy or girl in your class is shorter than you are.

3. Refer to the normal curves below for heights of all 11-year-old boys and of all 11-year-old girls.
    *   If a certain 11-year-old boy is 60 inches tall, estimate the percentage of all 11-year-old boys who are shorter than he is.
    *   One percent of all 11-year-old girls are taller than what height (approximately)?
    *   Use the normal curve for heights of *all* 11-year-old boys or girls to find the probability that a randomly chosen 11-year-old boy or girl is shorter than you are. [For example, if you are a boy who is 52 inches tall, you fall about a third of the way between 51.0 and 53.7. The area under the curve to the left of 52 would be about .02 + .03 = .05.]

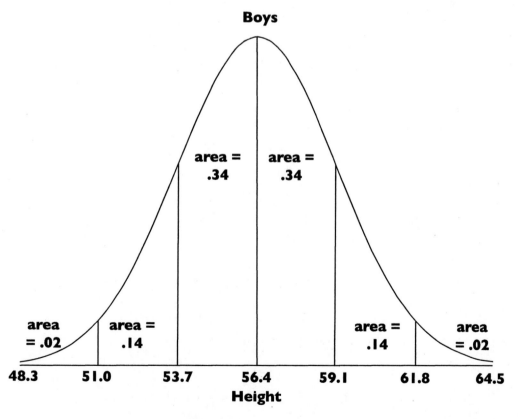

Boys

area = .34

area = .34

area = .02

area = .14

area = .14

area = .02

48.3   51.0   53.7   56.4   59.1   61.8   64.5

Height

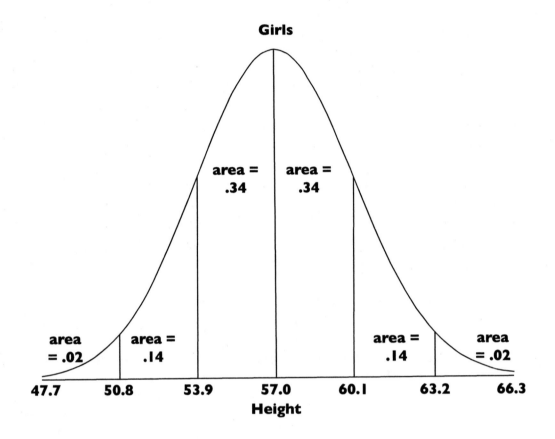

Girls

area =
.34

area =
.34

area
= .02

area =
.14

area =
.14

area
= .02

47.7   50.8   53.9   57.0   60.1   63.2   66.3

Height

Standard Deviation

Introduction

In general, the mean provides the best measure of center, but it does not give us a very complete picture of a distribution. To summarize a distribution adequately, we need to describe its spread. The best measure of spread—and the one which most naturally accompanies the mean—is the *standard deviation*.

Calculating Standard Deviation

Question   We have already seen that the heights of my family members are more spread out than the heights of class members. How can we actually measure the spread of a variable?

Answer   In general, the "best" way to measure the spread of a variable is to calculate the square root of the average squared distance from the mean.

The difference between each particular value of a variable and the mean value of that variable is called a **deviation**. Squaring each deviation prevents negative distances from canceling out positive distances and lets us report a higher spread for distributions with some very extreme values. Adding up all the squared deviations and then dividing by the number of values averages them out. Taking the square root "standardizes" the measurement so that it is in the same units as the variable itself. This is the procedure for calculating our preferred measure of spread.

Basic Rule	The **standard deviation**, which measures the spread of a variable, is calculated as follows: The differences of all the variable's values from its mean value are squared and added together. This sum is divided by the number of values used. Finally, the square root is taken. By its construction, the standard deviation measures the typical distance of values from their mean.

Example   Find the standard deviation of heights of my family members.

Solution   We make a chart of all 19 heights and the differences of each height from the mean height value 58. We square these differences, add them all together, and divide by 19. Finally, we take the square root.

Height	Difference from Mean	Squared Difference from Mean
29	29 - 58 = -29	-29 x -29 = 841
36	36 - 58 = -22	-22 x -22 = 484
45	45 - 58 = -13	-13 x -13 = 169
45	45 - 58 = -13	-13 x -13 = 169
46	46 - 48 = -12	-12 x -12 = 144
55	55 - 58 = -3	-3 x -3 = 9
56	56 - 58 = -2	-2 x -2 = 4
56	56 - 58 = -2	-2 x -2 = 4
57	57 - 58 = -1	-1 x -1 = 1
58	58 - 58 = 0	0 x 0 = 0
62	62 - 58 = 4	4 x 4 = 16
64	64 - 58 = 6	6 x 6 = 36
65	65 - 58 = 7	7 x 7 = 49
67	67 - 58 = 9	9 x 9 = 81
68	68 - 58 = 10	10 x 10 = 100
72	72 - 58 = 14	14 x 14 = 196
72	72 - 58 = 14	14 x 14 = 196
72	72 - 58 = 14	14 x 14 = 196
72	72 - 58 = 17	17 x 17 = 289

$$\frac{(841 + 484 + 169 + 169 + 144 + 9 + 4 + 4 + 1 + 0 + 16 + 36 + 49 + 81 + 100 + 196 + 196 + 196 + 289)}{19} = 157$$

[Our answer above has been rounded to the nearest whole number.] Since the square root of 157 is about 12.5 (in other words, 12.5 x 12.5 = 157), we have found that the standard deviation of my family members' heights is approximately 12.5 inches. Some heights are closer to the mean and some are further away, but we can say that 12.5 inches is the typical distance of a family member's height from the mean value, 58.

Question Which data set would have a larger standard deviation: the set of heights of my family members, or the set of heights of class members?

Answer Since heights of my family members are more spread out, they would have a larger standard deviation than heights of class members.

Basic Rule	In general, data sets that are more spread out have larger standard deviations than data sets that are less spread out.

Example Which of these two data sets has a larger standard deviation? Verify your answer by calculating and comparing both standard deviations:

Data Set #1:	2	4	6	8	10
Data Set #2:	4	5	6	7	8

Solution   The first data set is clearly more spread out than the second data set, so it must have a larger standard deviation. We can verify this as follows:

The mean of the first data set is $(2 + 4 + 6 + 8 + 10)/5 = 6$.
The mean of the second data set is $(4 + 5 + 6 + 7 + 8)/5 = 6$.

For the first set,
$$\frac{[(2 - 6) \times (2 - 6)] + [(4 - 6) \times (4 - 6)] + [(6 - 6) \times (6 - 6)] + [(8 - 6) \times (8 - 6)] + [(10 - 6) \times (10 - 6)]}{5}$$

$$= (16 + 4 + 0 + 4 + 16)/5 = 8.$$

Since $2.8 \times 2.8 = 8$ (approximately), the standard deviation of the first data set is about 2.8.

For the second set,
$$\frac{[(4 - 6) \times (4 - 6)] + [(5 - 6) \times (5 - 6)] + [(6 - 6) \times (6 - 6)] + [(7 - 6) \times (7 - 6)] + [(8 - 6) \times (8 - 6)]}{5}$$

$$= (4 + 1 + 0 + 1 + 4)/5 = 2.$$

Since $1.4 \times 1.4 = 2$ (approximately), the standard deviation of the second data set is about 1.4.

In fact, we can say that the first data set is twice as spread out as the second data set, because its standard deviation is twice as large.

Physical Interpretation of Standard Deviation

Question   What happens to a spinning figure skater when the skater brings his or her arms or legs in close to his or her body?

Answer   Pulling his or her arms or legs closer to his or her body makes the skater spin faster.

Basic Rule   Suppose an object is rotating about its balance point with a fixed amount of energy. The less spread out the object is, the faster it will rotate.

Chances Are ...

We have already equated the **mean** calculated by statisticians with the **balance point** calculated by physicists. Now we can take the comparison one step further: the same formula which is used by statisticians to calculate the **variance**, or squared standard deviation, is also used by physicists to calculate the **moment of inertia**, which measures how much energy is required to start or stop a body spinning about its balance point. For a fixed amount of energy, a body spins slowly if it has a large moment of inertia; it spins quickly if it has a small moment of inertia. Thus, a body which is spread further away from its center is harder to start (or stop) spinning than a body which is bunched up close to its center.

Example   First imagine that a yardstick is given a push to twirl about its middle. Next imagine that the yardstick is cut into three rulers which are stacked on top of one another and then given a push to twirl about the middle. If the push has equal force in both cases, which will twirl faster?

Solution   Since the stacked rulers are less spread out than the yardstick, they will twirl faster.

◆   ◆   ◆

Example   The following represent probability distributions of two different continuous random variables. Which of the two has a smaller standard deviation?

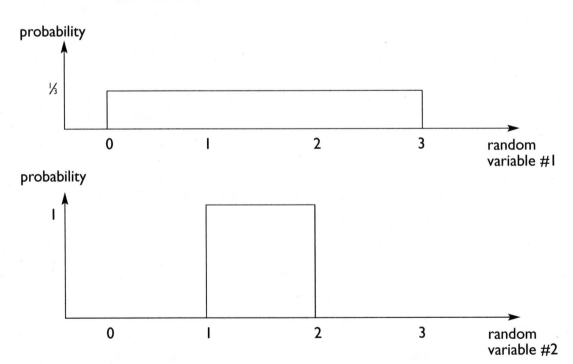

Solution   The second distribution is clearly less spread out, so it has a smaller standard deviation.

Experiment:
Changing Your "Standard Deviation"

Materials: swivel chair

1. Sitting in a swivel chair, use your feet to give yourself one push (as hard as you can) to spin around. As soon as you start spinning, extend your arms out as far as possible from your sides. Make a note of how many turns you are able to make until you come to a stop.

2. Starting at the same position as the first time, again give yourself one push as hard as you can. This time keep your arms as close as possible to your body. Is there a difference in how many turns you make this time? If so, in which position can you spin more? In which position do you have a larger "standard deviation"?

[Note: The order in which the experiment is performed with and without arms extended should not be allowed to bias your results. Ideally, this experiment should be performed by several students. Each student begins by flipping a coin. If a head appears, the first spin should be performed with arms extended and the second spin without. If a tail appears, the first spin should be performed without arms extended and the second spin with.]

Standard Deviation as a Statistic (Optional)

The distribution of my family members' heights was examined for its own sake, and we did not consider the values to be a sample taken from a larger group. The heights of class members, on the other hand, may be thought of as a sample taken from the distribution of heights of all children in the same age bracket. When we calculate the average squared distance from the mean of their heights, we may intend to use this as an estimate for the average squared distance from the mean of heights of all children in their age bracket. According to statistical theory, we will be guilty of bias (in this case, systematic underestimation) unless we divide by the number of values *minus one* when we calculate average squared distance from the mean.

Basic Rule	The **sample standard deviation** (a statistic) is calculated as follows: the differences of all the sample values from the sample mean value are squared and added together. This sum is divided by the number in the sample *minus one*. Then the square root is taken.

Example   Consider heights of class members to be a sample taken from the heights of all children in this age bracket. Calculate the sample standard deviation of their heights.

Solution   We first calculate all squared distances from the mean height, then divide their sum by the number of class members *minus one*, and finally take the square root.

Standard Deviation in a Normal Distribution

If we think of a boy being selected at random from *all* boys in a particular age bracket, then that boy's height is a continuous random variable. We cannot calculate the exact mean and standard deviation of this random variable, but there are ways to get good estimates for them. So, let us assume that we know the mean and the standard deviation of a random variable. If, in fact, the random variable follows a normal distribution (as do many random variables, including height) then its mean and standard deviation together tell us everything there is to know about its behavior. In particular, we can tell the probability that the random variable falls within a certain number of standard deviations from its mean.

Basic Rule   For any normal random variable,
the probability is 68% that the variable falls within one standard deviation of the mean;
the probability is 96% that the variable falls within two standard deviations of the mean;
the probability is almost 100% that the variable falls within three standard deviations of the mean.

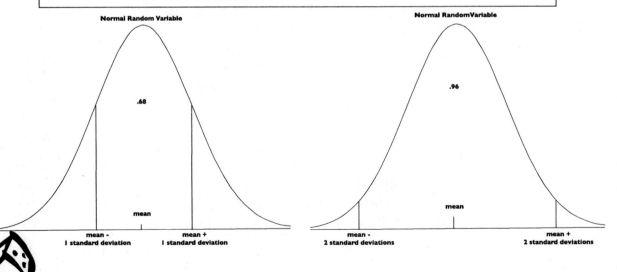

116

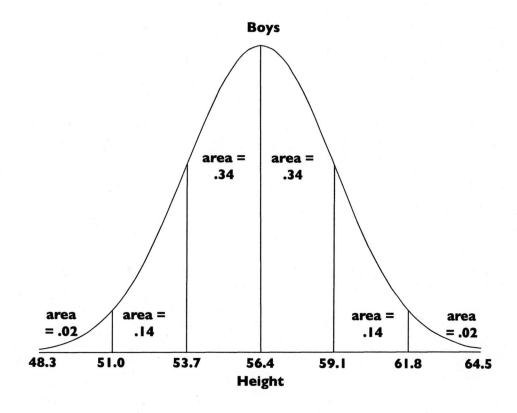

Example Height for 11-year-old boys is normally distributed, as shown below. Find the standard deviation of this distribution.

Solution According to the normal curve, the mean height of 11-year-old boys is 56.4 inches. Thirty-four percent of boys fall within 2.7 inches below the mean (down to 53.7 inches) and another 34% fall within 2.7 inches above the mean (up to 59.1 inches). Altogether, 68% fall within 2.7 inches on either side of the mean, so the stan-

dard deviation of the height distribution of 11-year-old boys is 2.7 inches.

[We can check further that 48% fall within 2 x 2.7 inches below the mean (down to 51.0) and 48% fall within 2 x 2.7 inches above the mean (up to 61.8), for a total of 96% within 2 standard deviations of the mean. Finally, 50% fall within 3 x 2.7 inches below 56.4 (down to 48.3) and the other 50% fall within 3 x 2.7 inches above 56.4 (up to 64.5), which accounts for 100% within three standard deviations.]

If we only know the mean and standard deviation of a normally distributed variable, then we know everything about its probability distribution.

Example   Height for 11-year-old girls is normally distributed with mean 57.0 inches and standard deviation 3.1 inches. Sketch its probability distribution.

Solution   We know the distribution is centered at 57.0 inches. Sixty-eight percent of the area under the normal curve falls within one standard deviation, or 3.1 inches, of 57.0—that is, between 53.9 and 60.1 inches. An additional 14% extends down another standard deviation, to 50.8 inches, and 14% up a standard deviation, to 63.2 inches. The remaining 2% of the area on either side extends a third standard deviation away from the mean, down to 47.7 inches and up to 66.3 inches.

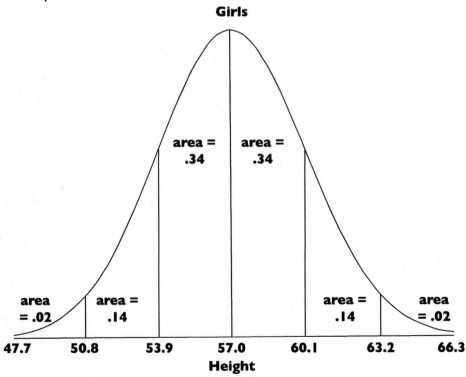

Distribution of Sample Mean and Sample Proportion

Suppose we intend to use the sample mean as an estimate for population mean. As a rule, only one sample is taken. For example, if we want to estimate the mean value of infinitely selected cards from a deck, we may take a sample of three cards and use its sample mean value as an estimate for population mean value. Or maybe we take one sample of five cards, and use its sample mean value as an estimate. Can we get a more accurate estimate with three cards or with five cards? For the purpose of determining the accuracy of our estimate, we will now consider values of the sample mean for *repeated* samples of the same size. We will determine how close to the population mean these sample mean values tend to fall, and think about the effects of different sample sizes. Similarly, we will look at how values of sample proportion behave for repeated samples, so we can decide how to get a better estimate when a single sample is used. Repeated samples may not make sense for practical purposes, but they help us to understand the theory behind sampling for good estimates.

Question   Suppose three cards are selected at random and replaced in an ordinary deck, and we use their sample mean value to predict the mean value of all the cards if we could go on picking forever. How good an estimate is the sample mean value for the population mean value (which we happen to know equals 7)?

Answer   We can examine how close the sample mean tends to be to the actual mean by repeating the selection of three cards enough times to see a pattern. For example, when I picked three cards 10 times, the 10 sample mean values were quite spread out, all the way from 4 up to 12. If one of these sample mean values were used as an estimate for the population mean, it may not do such a good job.

◆   ◆   ◆

Question   Now suppose *five* cards are selected at random and replaced in an ordinary deck, and we use their sample mean value to predict the mean value of all the cards if we could go on picking forever. How good an estimate is the sample mean for the actual mean?

Answer   When I picked *five* cards 10 times, the 10 sample mean values were not as spread out as the sample mean values when only *three* cards were picked 10 times. The lowest sample mean was about 5 and the highest was about 9. Now the sample mean tends to provide a closer, more accurate estimate for the population mean.

This simple experiment verifies a rule that can actually be proven using probability theory:

Basic Rule	In general, the sample mean from a larger sample gives a better estimate for the population mean than the sample mean from a smaller sample.

Example   Suppose we want to estimate mean height of all 11-year-olds. Which would provide a better estimate: (a) the mean height of two randomly selected 11-year-olds; or (b) the mean height of 20 randomly selected 11-year-olds?

Solution   A better estimate would be provided by (b) because the sample is larger.

Question   Suppose four cards are selected and replaced in an ordinary deck, and we use the sample proportion of diamonds to predict the proportion of diamonds which would be picked if we went on forever. How good an estimate is the sample proportion for the actual proportion (which we happen to know equals ¼)?

Answer   When I picked four cards 10 times, the sample proportion of diamonds was spread out between ¾ and ⅔.

Question   Now suppose *eight* cards are selected and replaced in an ordinary deck, and we use the sample proportion of diamonds to predict the population proportion of diamonds selected if we went on forever. How good an estimate is the sample proportion for the actual proportion?

Answer   When I picked *eight* cards 10 times, the sample proportion of diamonds was not so spread out: it ranged from ⅛ to ⅜.

Basic Rule	In general, the sample proportion from a larger sample gives a better estimate for the population proportion than the sample proportion from a smaller sample.

Example   If we want to get an estimate for the overall proportion of hits a baseball player gets at bat, which would provide a better estimate: (a) the sample proportion of hits in a single game; or (b) the sample proportion of hits in several games?

Solution   Measuring the sample proportion of hits from several games uses a larger sample, so this would give a better estimate.

The Central Limit Theorem

In practice, just one sample is taken in order to measure the *statistic* sample mean (or sample proportion). In theory, we may consider values of sample mean (or sample proportion) for repeated samples, in order to get an idea of how sample mean (or sample proportion) as a *variable* behaves. If the sample is chosen at random, then sample mean or sample proportion is actually a *random variable*.

Thinking of sample mean as a random variable, we can describe its behavior by summarizing its distribution, including mention of center, spread, and shape. What makes this all possible is the **Central Limit Theorem**, one of the most powerful tools in statistics.

1. First of all, it can be shown that the distribution of the sample mean (for all possible samples of a given size) is *centered* right at the mean value for the entire population from which the samples were taken! For example, if we could take all possible samples of a given size from the population of all 11-year-old boys, the distribution of sample mean height values would be centered at 56.4, because this is the population mean height. If we could take all possible samples of three cards from a deck, measuring the sample mean value each time, the distribution of all these sample mean values would be centered at 7.

2. As for the *spread* of the distribution of sample mean, we have already seen that sample mean values tend to be closer together for larger samples and spread farther apart for smaller samples. In fact, the Central Limit Theorem states explicitly that the standard deviation of the distribution of the sample mean decreases as the sample size increases. Thus, the sample mean height for all possible samples of 20 boys would have a distribution with a smaller standard deviation than the sample mean height for all possible samples of 10 boys. Likewise, the sample mean for all possible samples of five cards would have a distribution with a smaller standard deviation than the sample mean for all possible samples of three cards.

3. Finally, the Central Limit Theorem tells us about the *shape* of the distribution of the sample mean. If the underlying population has a normal distribution, then so does the sample mean! If the underlying population is not normally distributed, the sample mean still tends to follow an approximately normal distribution as long as the sample is large enough. For example, since heights of 11-year-old boys are normally distributed, the distribution of sample mean height for all possible samples of a given size is also normally distributed. Since the distribution of card values is not normal [think about how to represent this distribution with a histogram ...], the distribution of the sample mean for all samples of a given size would have a shape which is more or less normal, depending on how large the sample size is. For samples of five cards, the shape would be more normal than for samples of three cards.

121

Similar results hold for the distribution of sample proportion. For example, suppose 20% of the M&Ms™ in a huge bag are red, and we look at the distribution of sample proportion for all possible samples of a certain size. Then the distribution would be centered at .20, its standard deviation would be smaller for larger sample sizes, and the shape of the distribution would be approximately normal for a large enough sample size.

The Central Limit Theorem shows that in theory, not just in practice, larger samples provide better estimates. Furthermore, it enables statisticians to be very precise about how accurate an estimate the sample mean is for the population mean, and likewise for sample and population proportions.

Experiment:
Sample Mean Dice Totals
for Different Sample Sizes

Materials: 4 dice for each group

1. Rolling a die two times is like taking a sample of size 2 from an infinite number of single dice rolls. The sample mean of those two rolls could be used to estimate the population mean for an infinite number of dice rolls. We happen to know that the population mean must be 3.5. [Why?]

    Another two rolls would probably produce a different sample mean and, therefore, a different estimate for the population mean of 3.5. In general, how close would such an estimate be to the population mean? To get an idea of what kind of estimate we tend to obtain from a sample of two dice rolls, I took 10 such samples, as shown below, and calculated the sample mean for each of them.

Trial	2 rolls	Sample Means	Dist. from Mean	Squared Distance from Mean
1	1,2	1.5	1.5 - 3.6 = -2.1	-2.1 x -2.1 = 4.41
2	4,6	5.0		
3	3,3	3.0		
4	4,6	5.0		
5	2,2	2.0		
6	4,3	3.5		
7	2,5	3.5		
8	5,5	5.0		
9	1,5	3.0		
10	6,3	4.5		
		Mean = 3.6		Av. Sq. Dist. from Mean = 1.44

Square Root of Average Squared Distance from Mean = Standard Deviation = 1.2, since 1.2 x 1.2 = 1.44.

The sample mean is a random variable with a distribution of its own. We already know how to summarize the distribution of values of a variable: report their center, spread, and shape!

The center of the sample mean values is their mean, calculated to be 3.6. [Notice how close this is to the population mean, which we know to be 3.5.]

The spread of the sample mean values can be described using their standard deviation. Calculate it by completing the columns for "Distances from Mean" and "Squared Distances from Mean," then averaging those squared distances and finding the square root. Your answer should be close to 1.2.

As for the shape of the distribution, display all the sample mean values in a histogram and describe its shape.

2.   Rolling a die four times is like taking a sample of size 4 from an infinite number of dice rolls. The sample mean of four rolls could also be used to estimate the population mean. How close an estimate would this be? We'll examine the distribution of this sample mean by taking 10 such samples and observing their center, spread, and shape. Complete the table below, then display the sample mean values in a histogram.

Trial	4 rolls	Sample Means	Dist. from Mean	Squared Distance from Mean
1	3,5,3,1	3.00	3.00-3.70 = -.70	-.70 x -.70 = .4900
2	3,1,3,1	2.00		
3	4,5,6,5	5.00		
4	5,4,1,5	3.75		
5	4,6,4,2	4.00		
6	3,4,2,5	3.50		
7	5,1,4,5	3.75		
8	3,6,3,6	4.50		
9	5,4,2,1	3.00		
10	5,1,6,6	4.50		
		Mean = 3.70		Av. Sq. Dist. from Mean = .6975

Square Root of Average Squared Distance from Mean = Standard Deviation = .83, since .83 x .83 = .6975 (approximately).

3.   We have examined the distributions of sample means from samples of two dice rolls and from samples of four dice rolls. Both distributions are centered quite close to the population mean, but one has more spread than the other,

that is, a higher standard deviation. Which of the two sample means tends to produce a closer, more accurate estimate for population mean, the one from two dice rolls or the one from four dice rolls? In general, what sample size would produce a sample mean which provides a better estimate for population mean, size two or size four?

4. (Optional) Collect your own samples of two dice rolls and of four dice rolls to make charts similar to mine. Which sample mean has a distribution with a larger standard deviation?

Experiment:
Sample Proportion of Red M&Ms™
for Different Sample Sizes

Materials: several bags of mixed color (preferably small baking) M&Ms™, a teaspoon, a tablespoon, and a bowl

In this experiment, we will examine the distribution of the sample proportion of red M&Ms™ selected from a very large bowl of multicolored M&Ms™ for *different sample sizes*.

First, each student should randomly select one *teaspoonful* of M&Ms and calculate the sample proportion of red M&Ms™ in that teaspoonful. Results for all students should be compiled and displayed in a histogram.

Next, each student should randomly select one *tablespoonful* of M&Ms™ and calculate the sample proportion of red M&Ms™ in that tablespoonful. Results should be compiled and displayed in a histogram using the same scale as in the previous histogram so that a comparison may be made.

Summarize and compare the two distributions:
1. Where is each distribution centered (approximately)?
2. Compare the spreads of the two distributions.
3. How would you describe the shapes of the distributions?
4. If sample proportion of red M&Ms™ were to be used to estimate population proportion of red M&Ms™ in a large bag, which sample size tends to give a more accurate estimate, a teaspoonful or a tablespoonful?
5. (Optional) If the samples of size one teaspoonful or one tablespoonful were coming from a larger bag of M&Ms™, would this have affected your results?

Exercise:
Calculating Standard Deviations

Jessica	Kevin	Leah	Marvin
4	4	4	4
3	4	4	3.25
3	4	4	3.25
3	4	3	3.25
3	2	3	3.25
3	2	3	2.75
3	2	3	2.75
2	2	3	2.75
		3	2.75
		3	2
		3	
		0	

Point assignments from the grade reports of four students are given above (as for the experiment on page 108). Each of the students has an average of a "B," or 3 points. The standard deviation for Jessica's grades is calculated below as an example.

Gradepoints	Distances from Mean	Squared Distances from Mean
4	4 - 3 = 1	1 x 1 = 1
3	3 - 3 = 0	0 x 0 = 0
3	3 - 3 = 0	0 x 0 = 0
3	3 - 3 = 0	0 x 0 = 0
3	3 - 3 = 0	0 x 0 = 0
3	3 - 3 = 0	0 x 0 = 0
3	3 - 3 = 0	0 x 0 = 0
2	2 - 3 = -1	-1 x -1 = 1
Mean = 3		Average Squared Distance from Mean $= \frac{(1 + 0 + 0 + 0 + 0 + 0 + 0 + 1)}{8}$ $= \frac{2}{8} = \frac{1}{4}$

Square Root of Average Squared Distance from Mean = Standard Deviation = ½ since ½ x ½ = ¼.

1. Use the same process to calculate the standard deviation for the grades of Kevin, Leah, and Marvin (each separately).

125

2. Summarize the four students' performances. Did the histograms you drew for the experiment on page 108 suggest which standard deviations would be larger?

3. What would be the standard deviation for the grade distribution of a student who received all As?

Handling Data in Different Forms

Introduction

Data are produced when we record values of one or more variables (which may be numerical or categorical) for a set of people or things. By now we have learned various techniques for displaying and describing a set of data, depending on what form it takes. In order to take full advantage of all these available tools, it is first necessary to determine which are appropriate in a given setting.

Summarizing Distributions (Review)

Question   How can we summarize (display and describe) the heights of class members?

Answer   We could use a stemplot, boxplot, or histogram to display the distribution of heights and report their center and spread in order to describe them.

Basic Rule   If there is just one aspect of each person or thing that interests us, and it can be measured with a number value, then our data set consists of a list of values for *one numerical variable*. Such data can be displayed with a stemplot, boxplot, or histogram, and described with center and spread (either the five number summary or the mean and standard deviation).

Question   How can we summarize the genders of class members?

Answer   A very simple Venn diagram—just a circle in a rectangle—could be used to record the number of class members who are or are not male. The sample proportion of males and of females could be used as a numerical description.

127

| **Basic Rule** | If there is just one aspect of each person or thing that interests us, and it can be evaluated as either falling in or out of a certain category, then our data set consists of a list of values for *one categorical variable*. The data could be displayed with a Venn diagram (although we generally would not even bother in such a trivial case!) and described by reporting sample proportions. |

Question How can we summarize the heights and genders of class members?

Answer A back-to-back stemplot could be used to display heights of males and of females. To describe the data, we could report the center and spread of both male and female heights (preferably using mean and standard deviation).

| **Basic Rule** | Suppose there are two aspects of each person or thing that interest us. If we have *one numerical variable* and *one categorical variable with only two possible values*, the data can be displayed with a back-to-back stemplot. Within each of the two categories we can use center and spread to describe the data. [Side-by-side boxplots with the five number summary are also an option.] |

Question How can we summarize the heights and preferred T-shirt sizes (S, M, L, or XL) of class members?

Answer In this case we could use four side-by-side boxplots to display the data, and we could describe the distribution of heights in each shirt-size category by reporting the five number summary or the mean and standard deviation.

| **Basic Rule** | Suppose there are two aspects of each person or thing that interest us. If we have *one numerical variable* and *one categorical variable with more than two possible values*, then the data may be displayed with a side-by-side boxplot. As a description we can report either the five number summary for each category, or the mean and standard deviation. |

Question How can we summarize class members' gender and whether or not they participated in organized sports outside of school this year?

Answer As a display technique, we could use either a Venn diagram with two circles or a two-way table (e.g. two columns for male or

female, two rows for sports or no sports). One way to describe the data would be to report the count or the proportion of class members in each of the four category combinations.

[Another possibility would be to use "marginal distributions" which concentrate on one variable at a time: we could report the proportion of students who are male and who are female, or the proportion who did participate in sports and the proportion who did not. In different contexts you may be interested in different marginal distributions.]

Basic Rule	Suppose there are two aspects of each person or thing that interest us, namely *two categorical variables with only two possibilities for each category.* The data may be displayed with a Venn diagram or a two-way table. Depending on the situation, we can use counts or the appropriate proportions to describe the data.

Question How can we summarize class members' gender and their preferred T-shirt size?

Answer A two-way table (e.g. two columns for male or female, four rows for each of four T-shirt sizes) can be used to display the data. Appropriate counts or proportions may be used to describe the data.

Basic Rule	Suppose there are two aspects of each person or thing that interest us: *two categorical variables with more than two possibilities in one or both categories.* Then the data may be displayed with a two-way table. Depending on the situation, appropriate counts or proportions may be reported to describe the data.

Question How can we summarize class members' gender, whether or not they participated in organized sports this year, and whether or not they took musical instrument lessons?

Answer A Venn diagram with three circles, or two two-way tables (say, one for males and one for females), would be possible display techniques. To describe the data, appropriate counts or proportions could be reported.

Basic Rule	In a situation that has *three categorical variables with only two possibilities in each category,* a Venn diagram with three circles or two two-way tables may be used to display the data, and appropriate counts or proportions may be used to describe it.

Summarizing Relationships

Up until now, we have dealt with situations involving no more than one numerical variable. As soon as a second numerical variable enters the picture, the issue of its relationship with the first numerical variable should be addressed. The study of relationships between numerical variables is one of statisticians' most important tasks.

Question   How can we summarize class members' heights and foot lengths?

Answer   As a display, a **scatterplot** could be used: starting at 0 height and 0 foot length, count over to the right for each student's foot length, count up for his or her height (or vice versa), and mark the point with a dot. [Alternatively, your plot could be started at some minimum height and foot length values, for example at 45 inches and 6 inches instead of at 0 inches for both.] The completed scatterplot should contain a dot for each student. If several students have the same height and foot length combination, the number of students with those values could take the place of the single dot.

A description of the data should give an idea of what kind of relationship (if any) there appears to be between the numerical variables height and foot length.

First, do large height values tend to occur with large foot length values, and small height values with small foot length values? If so, the relationship is **positive**. Or is there a **negative** relationship, where large height values occur with small foot length values, and vice versa? Common sense tells us that this particular relationship should be positive, not negative.

Next, we observe whether the scatterplot points seem to be falling roughly along a straight line. This is important because there are very efficient ways of analyzing straight-line relationships in statistics. Within a specific age group, we can expect the relationship between height and foot length to be fairly **linear**, not **curved**.

Finally, we should make a note as to how strong a relationship we see between the two variables. This is usually not such a simple judgment call. An example will be done later so that we can compare scatterplots for **strong** and **weak** relationships. [Statisticians also can make use of a complicated formula to calculate the **correlation**, a number between 0 and 1 which measures the strength of any linear relationship between two numerical variables.]

Because of the way human beings grow, it is reasonable to anticipate a fairly strong relationship between class members' heights and foot lengths.

Basic Rule	If there are two number values of interest for each person or thing, then we need to summarize the relationship between *two numerical variables*. A **scatterplot** serves as the best display technique. To describe the data, we should pay attention to three aspects of this relationship (that is, if there is a relationship at all): **direction** (positive or negative), **form** (linear or not), and **strength** (strong or weak).

Question	How can we summarize class members' heights, foot lengths, and genders?

Answer	A scatterplot could be drawn with an "M" marking the point for each male student's height and foot length, and an "F" for each female. The description could include details about the direction, form, and strength of the relationship for males and females separately.

Basic Rule	If the data consist of *two numerical variables* and *one categorical variable*, a labeled scatterplot can be used to display the relationship for both categories. Description of direction, form, and strength of the relationship can include details about both categories.

Experiment:
Relationships of Other Variables to Age

Consider the numerical variable "Age" for students between 10 and 18 years old.

1. Think of another numerical variable that you believe would have a **positive** relationship with "Age" between 10 and 18 years. Write the variable name next to the vertical axis arrow. Then draw in points on the scatterplot below to show roughly what kind of pattern you would expect to see if a large group of students in this age bracket were surveyed.

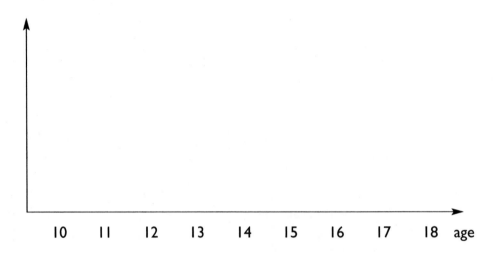

| 10 | 11 | 12 | 13 | 14 | 15 | 16 | 17 | 18 | age |

2. Think of another numerical variable which would have a **negative** relationship with "Age"; write in its name and fill in another scatterplot.

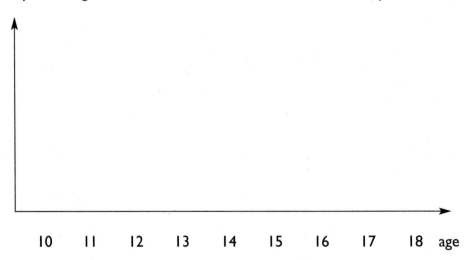

| 10 | 11 | 12 | 13 | 14 | 15 | 16 | 17 | 18 | age |

3. Think of a numerical variable which is likely to have **no** relationship with "Age"; write in its name and fill in the third scatterplot as it might appear.

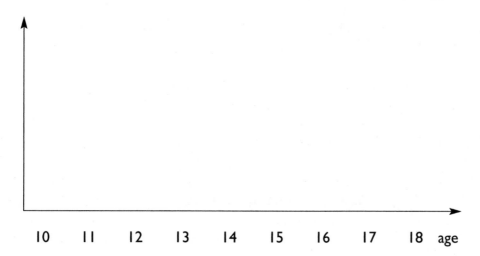

10	11	12	13	14	15	16	17	18	age

Experiment:
Relationships in Card Selections

Materials: decks of cards

A hand of 10 cards was dealt at random from an ordinary deck of cards. The numbers of red cards, hearts, black cards, and face cards (jack, queen, or king) were recorded and the cards were replaced and shuffled. Another hand of 10 cards was randomly dealt, and so on, for a total of 15 trials. Results are shown below:

Trial #	Reds	Hearts	Blacks	Face Cards
1.	6	4	4	3
2.	4	2	6	2
3.	4	2	6	3
4.	2	1	8	2
5.	6	2	4	3
6.	6	5	4	2
7.	7	3	3	2
8.	8	4	2	1
9.	7	3	3	2
10.	5	1	5	1
11.	8	5	2	2
12.	5	2	5	2
13.	6	3	4	2
14.	4	3	6	2
15.	7	3	3	1

1. Summarize the relationship between the number of red cards and the number of hearts.

2. Summarize the relationship between the number of red cards and the number of black cards.

3. Summarize the relationship between the number of red cards and the number of face cards.

Solution
1. The relationship between hearts and red cards is *positive*: a hand with more red cards tends to have more hearts, and a hand with fewer red cards tends to have fewer hearts. We can call the relationship *linear* because a straight line, rather than a curve, would fit the points best. The relationship is on the *weak* side, because random card selections naturally give various results.

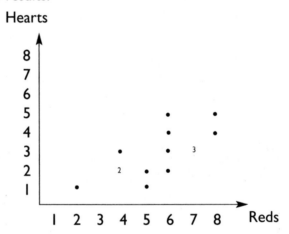

2. The relationship between black cards and red cards is *negative*: more red cards are dealt with fewer black cards, and fewer red cards are dealt with more black cards. The relationship is perfectly *linear*, and as *strong* as it can possibly be: the number of red cards dealt exactly determines how many black cards were dealt.

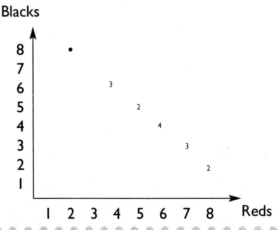

3. Our scatterplot shows *no relationship* between face cards and red cards: the points on the plot are scattered about with no visible pattern. Since there are just as many black face cards as there are red face cards, dealing more or fewer red cards will not affect the number of face cards dealt.

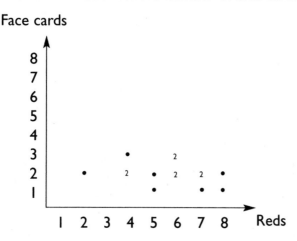

Face cards

4. Perform this experiment yourself, selecting 15 random hands of 10 cards. Draw and analyze scatterplots as is done in numbers 1, 2, and 3 above.

5. Without any data for the number of spades selected, what kind of relationship (if any) would you expect to see between the number of red cards and the number of spades?

Exercise:
Summarizing Data

Materials: completed surveys

Use the results of a class survey to summarize (display and describe) the data for the following variables:

1. height
2. gender
3. height and gender
4. height and preferred t-shirt size
5. gender and participation in a sports team
6. gender and preferred t-shirt size
7. gender, participation in a sports team, and music lessons
8. height and foot length (barefoot, to the nearest quarter inch)
9. height, foot length, and gender

Chances Are ...

Survey #	Height	Gender	T-shirt Size	Sports?	Music Lessons?	Foot length
1						
2						
3						
4						
5						
6						
7						
8						
9						
10						
11						
12						
13						
14						
15						
16						
17						
18						
19						
20						

Statistical Inference II (Optional)

Introduction

One of the most important goals in statistics is to use a sample statistic to estimate a population parameter. For instance, if we take a random sample of 100 adult males and find their sample mean height to be 70 inches, then we agree that 70 inches would be a "good" estimate for the population mean height of all adult males. But how good is "good"?

We have seen from our experiments and from the Central Limit Theorem that estimates from smaller samples are not as good as those from larger samples. For example, if we only measured the height of one adult male, this estimate would tend to be "less accurate" than the one for 100 adult males. But how much less accurate is "less accurate"?

When using sample mean to estimate population mean, our work is not finished until we can qualify the accuracy of this estimate using precise numbers rather than vague words.

Using statistics to estimate parameters is called *statistical inference* because we are inferring something about the whole population from the sample taken. One form of statistical inference is called a **confidence interval**, in which we state how confident we are that the unknown parameter falls within a certain interval. A typical confidence interval statement would have the following form: "We are 96% confident that the mean height of all adult males is between 69.5 and 70.5 inches."

The other form of statistical inference, called **hypothesis testing** (see Statistical Inference I), involves checking for evidence against the claim that a parameter takes on a certain value. For example, we might say, "If the population mean height of adult males were 71 inches, then the probability that a sample of size 100 would produce a sample mean as low as 70 inches is almost zero."

Confidence intervals are frequently seen in news reports, and hypothesis tests are common in academic research. Unfortunately, both methods are often misunderstood by those who would like to use them.

Confidence Intervals

We begin by restating our rule for the probability distribution of a normal random variable from page 116:

For any normal random variable,

the probability is 68% that the variable falls within one standard deviation of the mean;

the probability is 96% that the variable falls within 2 standard deviations of the mean; and

the probability is almost 100% that the variable falls within three standard deviations of the mean.

Question   Suppose we know that heights of adult males are normally distributed with standard deviation 2.5, but we do not know the population mean height. What does our rule tell us about the height of a randomly chosen adult male?

Answer   If an adult male is chosen at random, the probability is 68% that his height will fall within 2.5 inches of the population mean height;

the probability is 96% that his height will fall within 2 x 2.5 inches of the population mean height;

the probability is almost 100% that his height will fall within 3 x 2.5 inches of the population mean height.

In this section, we will frequently concentrate on probabilities or levels of confidence for a variable falling within 1, 2, or 3 standard deviations of the mean. Our results can be summarized by sketching the appropriate intervals:

Probability Intervals for Height of Randomly Chosen Individual

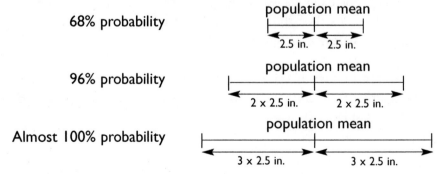

The statements on the previous page tell where an individual height tends to be. Our final goal, however, is to tell where the unknown population mean height tends to be.

Question How can we re-express the statements on the previous page to tell where the population mean height of adult males tends to be relative to a randomly chosen individual height?

Answer Observe that if I am within one foot's distance of you, then you must be within one foot's distance of me! This kind of reasoning justifies our switching the roles of population mean height and randomly chosen individual height in the above statements. However, we must be careful because the height of a randomly chosen individual is a random variable, whereas the population mean height is not. Although it may be unknown, population mean height is fixed and does not vary. Therefore, we cannot talk about the *probability* of its falling in a certain range. Instead, we talk about how *confident* we are that it falls in a certain range.

Our confidence level is 68% that the population mean height is within 2.5 inches of the height of a randomly chosen adult male;

Our confidence level is 96% that the population mean height is within 2 x 2.5 inches of the height of a randomly chosen adult male;

Our confidence level is almost 100% that the population mean height is within 3 x 2.5 inches of the height of a randomly chosen adult male.

Confidence Intervals for Population Mean Height

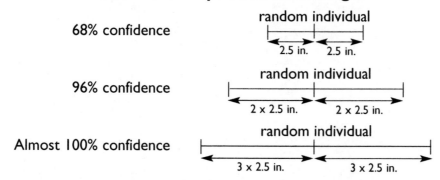

Now, suppose a particular adult male is chosen at random, and his height is measured to be 70 inches. Then our confidence intervals would be as follows.

We are 68% confident that the population mean height is within 2.5 inches of 70 inches (i.e. between 67.5 and 72.5 inches).

We are 96% confident that the population mean height is within 2 x 2.5 inches of 70 inches (i.e. between 65 and 75 inches).

We are almost 100% confident that the population mean height is within 3 x 2.5 inches of 70 inches (i.e. between 62.5 and 77.5 inches).

Confidence Intervals for Population Mean Height

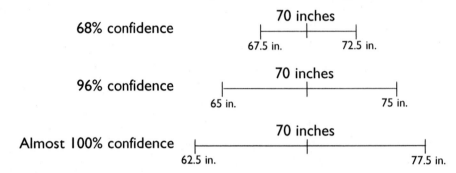

68% confidence

70 inches

67.5 in.   72.5 in.

96% confidence

70 inches

65 in.   75 in.

Almost 100% confidence

70 inches

62.5 in.   77.5 in.

The above confidence interval statements, although accurate, are not especially useful, because the intervals of heights are too wide. Being 96% confident that the population mean height is anywhere between 65 and 75 inches just doesn't tell us much. The key to making precise estimates in statistics is to take samples and use the Central Limit Theorem.

We stated on page 121 that if we take samples from a normal population, then the sample mean is a random variable which also has a normal distribution, centered at the mean value for the underlying population. For example, we know that heights of 11-year-old girls are normally distributed with mean 57 inches, as shown in the picture below.

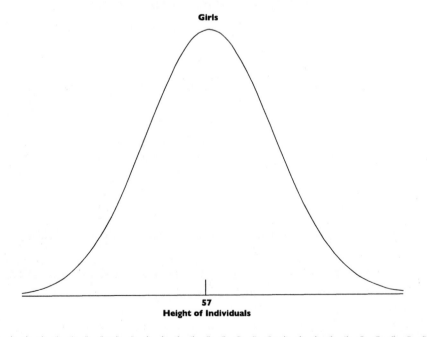

Girls

57
Height of Individuals

If we took many samples of a certain size from the population of 11-year-old girls and displayed all the sample mean values in a histogram, its center would be at 57 inches (approximately) and its shape would be normal (approximately):

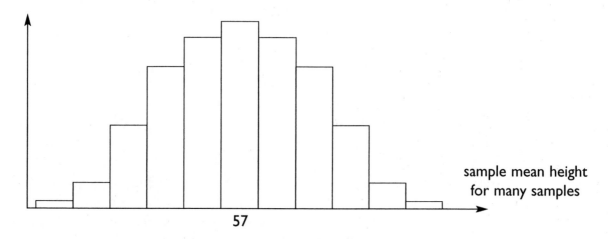

sample mean height for many samples

57

If we could take *all* possible samples of a given size from an infinite population of 11-year-old girls, their sample mean, as a continuous random variable, would have a normal distribution, centered at 57 inches:

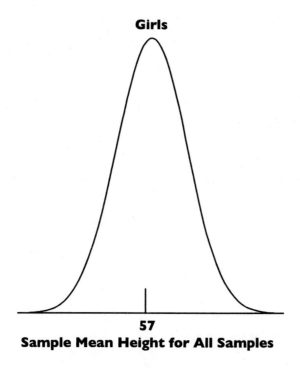

Girls

57
Sample Mean Height for All Samples

What about the *spread* of the variable "sample mean"? On page 121, we explained that the spread would be large for small samples and small for large samples. In fact, the Central Limit Theorem is more specific than that: the standard deviation of the distribution of the sample mean equals the standard deviation of the original population, divided by the square root of the sample size! This is what makes

random sampling such a powerful tool: it enables us to use statistics which come close to the parameters they are estimating and lets us calculate just how close they are.

Basic Rule **(from the Central Limit Theorem)** Suppose a random sample of a certain size is taken from a normal population. Then

1. the distribution of the sample mean is centered at the population mean;
2. its standard deviation is the population standard deviation divided by the square root of the sample size; and
3. its shape is normal.

Question Again, suppose we know that heights of adult males are normally distributed with standard deviation 2.5 inches and unknown population mean. For a random sample of 100 men, what does the Central Limit Theorem tell us about the distribution of sample mean height?

Answer
1. Sample mean height will be centered at the unknown population mean height.
2. Since the population standard deviation is 2.5 inches and the square root of the sample size is 10 [because 10 x 10 = 100], the standard deviation of the sample mean is $^{2.5}/_{10}$ = 0.25 inches.
3. Since height is normally distributed, sample mean height must also be normally distributed.

◆   ◆   ◆

Question What does our basic rule from page 121 tell us about the sample mean height from a random sample of 100 adult males?

Answer For a random sample of 100 adult males,

the probability is 68% that sample mean height is within 0.25 inches of population mean height;

the probability is 96% that sample mean height is within 0.50 inches of population mean height;

the probability is almost 100% that sample mean height is within 0.75 inches of population mean height.

Probability Intervals for Sample Mean Height

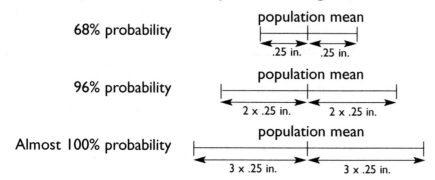

68% probability

population mean

.25 in.   .25 in.

96% probability

population mean

2 x .25 in.   2 x .25 in.

Almost 100% probability

population mean

3 x .25 in.   3 x .25 in.

The step we are about to take is a crucial one because it is the step which provides the connection from probability to statistics. Probability theory starts with a known parameter (like population mean) and tells us how the associated statistic (like sample mean) behaves. Statistical inference starts with a statistic (like sample mean) and uses it to make a statement about a parameter (like population mean).

As on page 138, we would like our statements to be about the unknown population mean height, so we make a similar argument: If sample mean height is within a certain distance of population mean height, then population mean height must be within that same distance of sample mean height! Again, we must use the word "confidence" because probability is for random variables, not for fixed values like population mean.

Question   How can we re-express the above statements to describe where the unknown population mean height for adult males tends to be relative to sample mean height?

Answer   We are 68% confident that population mean height is within 0.25 inches of sample mean height;

we are 96% confident that population mean height is within 0.50 inches of sample mean height;

we are almost 100% confident that population mean height is within 0.75 inches of sample mean height.

When we think about the behavior of sample mean if random samples were taken repeatedly, then sample mean is a random variable. When we actually take one sample and measure the value of the sample mean for that sample, then sample mean is a statistic. Once we have measured the statistic "sample mean," we can use confidence intervals to be very specific about where the population mean should be.

Question   Now suppose one sample of 100 adult males is chosen at random, and its sample mean height (a statistic) is measured to be 70 inches. What can we say about the population mean height relative to this particular sample mean?

Answer   We are 68% confident that population mean height is within 0.25 inches of 70 inches (i.e. between 69.75 and 70.25 inches).

We are 96% confident that population mean height is within 0.50 inches of 70 inches (i.e. between 69.50 and 70.50 inches).

We are almost 100% confident that population mean height is within 0.75 inches of 70 inches (i.e. between 69.25 and 70.75 inches).

Confidence Intervals for Population Mean Height

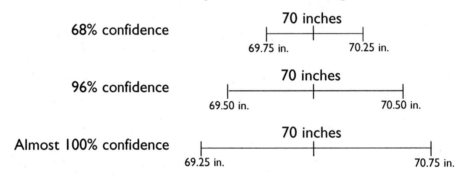

Observe how much more precise we can be about population mean using sample mean from a sample of size 100 as our estimate, instead of a single height value. If we only measure a single randomly chosen adult male height to be 70 inches, we can be 96% confident that population mean height is between 65 and 75 inches. If we measure sample mean height from 100 randomly chosen adult males to be 70 inches, then we can be 96% confident that population mean height is between 69.5 and 70.5 inches.

**96% Confidence Interval for Population Mean Height
Based on Random Sample of Size 1**

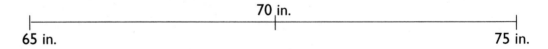

**96% Confidence Interval for Population Mean Height
Based on Random Sample of Size 100**

70 in.
69.5 in.   70.5 in.

Basic Rule   Suppose a random sample is taken from a normal population with unknown mean and known standard deviation, and the sample mean is measured. Then we can calculate the following **confidence intervals**:

We are 68% confident that the population mean is within

$$\frac{\text{one standard deviation}}{\text{square root of sample size}}$$

of the sample mean.

We are 96% confident that the population mean is within

$$\frac{\text{two standard deviations}}{\text{square root of sample size}}$$

of the sample mean.

We are almost 100% confident that the population mean is within

$$\frac{\text{three standard deviations}}{\text{square root of sample size}}$$

of the sample mean.

Example   Scholastic Aptitude Test scores in a population are normally distributed with standard deviation 100 points. Suppose a random sample of 25 scores is taken and the sample mean score is measured to be 500 points. Find three confidence intervals for the unknown population mean score.

Solution   Since the square root of 25 is 5, we can use our basic rule above to calculate confidence intervals as follows:

We are 68% confident the population mean score is within $\frac{100}{5} = 20$ points of 500 points, i.e. between 480 and 520 points.

We are 96% confident the population mean score is within $\frac{200}{5} = 40$ points of 500 points, i.e. between 460 and 540 points.

We are almost 100% confident the population mean score is within $\frac{300}{5} = 60$ points of 500 points, i.e. between 440 and 560 points.

Confidence Intervals for Population Mean Test Score

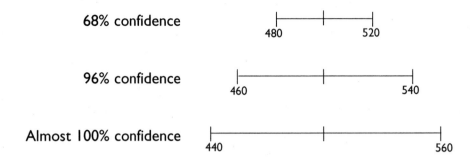

68% confidence   480   520

96% confidence   460   540

Almost 100% confidence   440   560

[Note: It may seem totally impractical to assume in a confidence interval or hypothesis testing problem that the population mean is unknown whereas the standard deviation is known. However, the methods used for problems with known standard deviation are quite similar to those used for problems with unknown standard deviation. Statistics courses at a more advanced level simply develop these same ideas a bit further to deal with unknown standard deviations. In this case, confidence intervals are calculated from a "t distribution" instead of from a normal distribution.

Another assumption that may seem to limit the applicability of our theory is that the underlying population be normally distributed. In fact, the Central Limit Theorem allows us to consider the sample mean to be approximately normally distributed even if the population is not—all that is necessary is that the sample be reasonably large.]

Hypothesis Testing

Besides confidence intervals, statisticians use statistics (like sample mean) to draw conclusions about parameters (like population mean) in another context: hypothesis testing. We start with a background quite similar to that for confidence intervals: a random sample is taken from a normal population with unknown mean and known standard deviation, and the sample mean is measured. Unlike confidence intervals, though, we include mention of a value which may or may not be the unknown population mean. Our mission is to decide whether or not it is reasonable, under the circumstances, to assume that the population mean takes on this proposed value. The "hypothesis" is that the population mean does in fact equal the value we suggest. The "testing" is really running a check to see if the laws of probability cast serious doubts upon our hypothesis.

Suppose we know that heights of adult males are normally distributed with standard deviation 2.5 inches. Population mean height is unknown, but someone claims it is 71 inches. Suspecting that the population mean is actually less than 71

inches, we run a test on the hypothesis: we get a random sample of 100 adult male heights and find their sample mean to be 70 inches.

Question   What is the hypothesis to be tested?

Answer   The hypothesis to be tested is that population mean height for adult males is 71 inches.

◆   ◆   ◆

Question   If the hypothesis is true, then what can we say about the likelihood of a sample of 100 men having a sample mean height of 71 inches?

Answer   If the hypothesis is true, then the sample mean, which is normally distributed, has mean 71 inches. Its standard deviation is 2.5 inches divided by the square root of 100, or $^{2.5}/_{10}$ = .25 inches. Thus, we have the following probability intervals:

Probability Intervals for Sample Mean Height

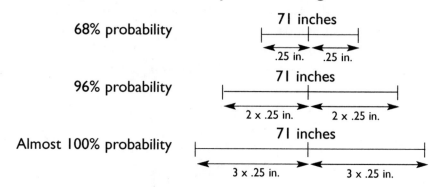

The distribution of sample mean would look like this:

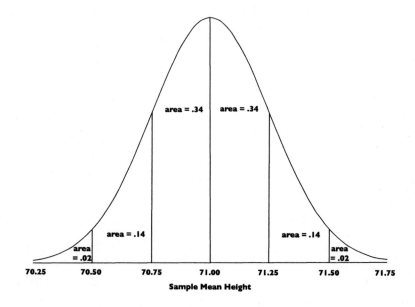

Since the area under this curve to the left of 70 inches is almost zero, it is almost impossible for a sample of size 100 coming from a population with mean 71 inches to have a sample mean as low as 70 inches. The sample mean of 70 inches has been measured and we cannot argue with it, but we can conclude that it is almost impossible for our sample to have come from a population with mean 71 inches. Therefore, we reject the claim that the population mean is 71 inches. It looks like the population mean must be smaller than 71 inches.

Example   Scholastic Aptitude Test scores from a population are normally distributed with standard deviation 100 points. The population mean score is unknown, but someone believes it to be 480 points. However, we have reason to suspect that it is more than 480 points. If a random sample of 25 scores has sample mean 500 points, does this give us good reason to doubt that the population mean is 480 points as claimed?

Solution   If the population mean is 480 points as claimed, then sample mean is also centered at 480. Since the population standard deviation is 100 points and the sample size is 25 (whose square root is 5), the sample mean has standard deviation $^{100}/_{5} = 20$ points. Also, since the population is normally distributed, sample mean must also be normally distributed. Thus, we have the following probability intervals:

Probability Intervals for Sample Mean Score

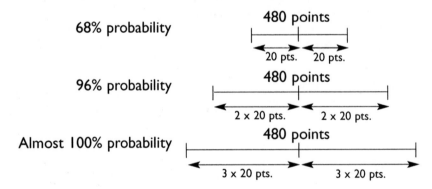

Under the assumption that the population mean is 480, the distribution of the sample mean must look like this:

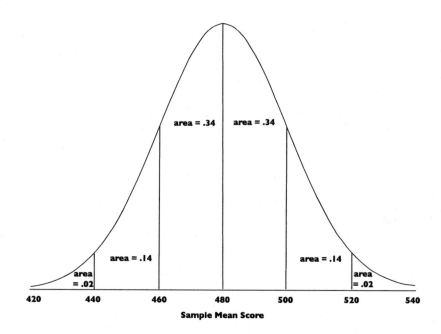

The probability that the sample mean would be at least as high as 500 equals the area under the curve to the right of 500, which is .14 + .02 = .16. If the population mean were 480 as claimed, then the probability that a sample of size 100 would have a sample mean value of 500 or more is .16—not so unlikely. The fact that the sample of 100 scores has mean 500 points does not provide very strong evidence against the claim that the population mean is 480, so in the end we accept the claim.

Example   It is known that scores on a certain IQ test are normally distributed with standard deviation equal to 20. The population mean score is unknown, but someone claims it is 100. Suppose we have doubts about the general accuracy of this claim, but wouldn't want to commit ourselves in advance as to whether we believe the true population mean score is actually higher or lower than 100. A random sample of 16 scores is taken, and their sample mean is measured to be 110. Does this provide strong evidence against the claim that the population mean is 100?

Solution   If the population mean is 100, then the distribution of the sample mean must be centered at 100. Since the population standard deviation is 20, and the sample size is 16 (whose square root is 4), the sample mean must have standard deviation $^{20}/_4$ = 5. Since the population is normally distributed, the sample mean must also be normally distributed. Here are the probability intervals for sample mean IQ:

Probability Intervals for Sample Mean IQ

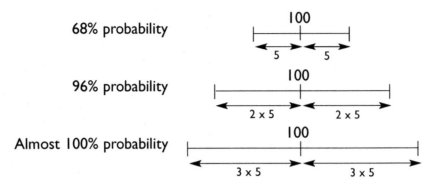

The probability distribution for sample mean IQ can be represented by the normal curve below:

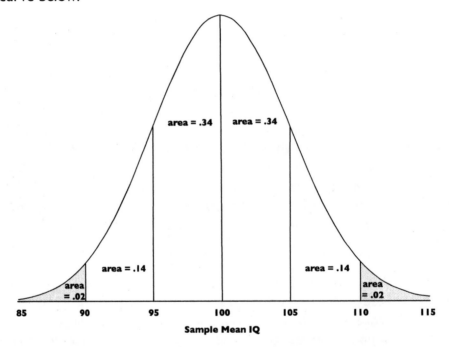

The shaded area shows the probability that the sample mean would take on a value as far away from the population mean as 110 is: that is, as high as 110 or as low as 90, which is 10 points away from the population mean in the other direction. The probability of getting a sample mean at least this far away from the population mean *in either direction* is .02 + .02 = .04. If an event can occur by chance alone only 4% of the time, we have some reason to doubt that it has actually occurred. The measured mean of 110 from our random sample of 16 scores gives us moderately strong evidence against the claim that the population mean score is 100.

Experiment:
Statistical Inference for Head Size

Materials: tape measure or string and ruler

1. **(Confidence Intervals)** The circumference of adult males' heads is known to vary normally with standard deviation approximately one inch.

    Measure a "random" sample of adult male head circumferences to the nearest quarter inch, avoiding multiple family members or any other circumstances which may lead to bias. [Students could each contribute a measurement to the sample, taken prior to the day of the experiment, using a tape measure or a piece of string and a ruler. For ease of calculation, you may want to use a sample of size 4, 9, or 16.] Find the sample mean head circumference and use this to construct three confidence intervals for the mean head circumference of all adult males.

2. **(Hypothesis Testing)** The circumference of adult males' heads is known to vary normally with standard deviation approximately one inch. The mean circumference is unknown to you, but I claim it to be 25 inches.
    *   Do you suspect the true mean is actually less than, or more than, or simply different from 25 inches?
    *   Can you use a sample of just one head circumference as statistical evidence against my claim? If my claim were correct, then the sample mean (which is normally distributed) would be centered at 25 inches. What is its standard deviation for a sample of size one? Sketch the distribution of the sample mean, under my claim, and find the approximate probability of getting a sample circumference at least as extreme as the one you measured. Does this provide you with convincing statistical evidence against my claim that mean head circumference is 25 inches?
    *   Now use a sample of several head circumferences to try to disprove my claim that the mean circumference is 25 inches. The sample mean would still be normal and centered at 25 inches. What is its standard deviation for the sample size you are using? Sketch the sample mean's distribution and find the approximate probability that it would take on a value such as you have measured. Does this provide you with convincing statistical evidence against my claim?
    *   Compare the evidence you gathered for one and for several head circumferences, and use this to explain the role of sample size in hypothesis testing.
    *   Can you think of any drawbacks to taking a very large sample? [Consider your experience with measuring head circumferences ...]

151

Appendix

Note to Instructor

This text was originally written for and used by a class of 20 gifted school students, entering grades four through six, who had signed up for a two-week (30 hours total) C-MITES (Carnegie-Mellon Investigation of Talented Elementary Students) summer course in probability and statistics. Because of time constraints, part of the material was dispensed with, such as some of the lengthier examples, and most of the optional sections.

A typical three-hour day began with a student coming forward to present his or her solution to the previous day's exercise. Presentation of alternate solutions was welcomed. Next, the day's text material was covered by the instructor using a chalkboard; after reasoning through each initial question and answer, then generalizing to the accompanying basic rule, students immediately applied what they had learned by working through each subsequent example, either in their heads or on paper. A 15-minute break usually took place about two-thirds of the way through this period. Once the theory was covered (roughly 10 pages' worth), students had a chance to put it into practice with the day's experiment. This was done in "tables" of four students whose desks had been pushed together for a group effort. Finally, one group of four was selected to present its findings to the rest of the class. Some students, when they'd completed their experiment early, opted to get a head start on the day's assigned exercise. Others occupied themselves with the cards, dice, dominoes, and roulette wheel which were also available before class started and during the breaks.

A contest "of, by, and for the students" took place on the very last day. From the first day, students had been encouraged to make up, solve, and hand in their own problems, similar to those done in class. These were compiled, with minor adjustments, and handed out to each table for a competitive/cooperative effort. Small prizes and certificates were awarded afterward to all the students for their individual contributions to the success of the course.

There are, of course, other possible structurings of the curriculum. Three or four one-hour periods could be covered each week during a semester, with the experiments capping off the end of each week. Alternatively, a three-hour unit could be covered once a week during a school semester—for instance, as a full morning's or afternoon's instruction at a gifted center. High school students could work through the material as an independent study course, either individually or in small groups, with regular monitoring by a teacher. This book can also serve as a

back-up for college students who, for their own personal use, need a reference with more elementary explanations than those offered by their course text.

As for the students' contributed problems, these could be included either as an end-of-course contest or as part of exams, if the latter are required.

For more information about C-MITES (Carnegie Mellon Institute for Talented Elementary Students), contact

Dr. Ann Lupkowski-Shoplik
Director, C-MITES
4902 Forbes Avenue, Box 6261
Carnegie Mellon University
Pittsburgh, PA 15213-3890
Tel: 412-268-1629
Fax: 412-268-8297

Sample Syllabus

Lesson 1
    Text pages 1–13
        1.   Experiment: Counting Color Arrangements
        2.   Combinatorics: Several-Stage Processes, Permutations, Combinations
        3.   Complete color experiment; Students present results
        4.   Fill out surveys
    Lesson 1–2 Exercise: Pascal's Triangle

Lesson 2
    Text pages 19–24
        1.   Student presents Exercise
        2.   Introduction to Probability
        3.   Experiment: Probabilities vs. Actual Proportions
        4.   Students present results
    Lesson 2–3 Exercise: Birthdays

Lesson 3
    Text pages 29–37
        1.   Student presents Exercise
        2.   Venn Diagrams; Independence/Dependence; Conditional Probability
        3.   Experiment: Venn Diagrams

4.   Students present results
Lesson 3–4 Exercise: Venn Diagrams

Lesson 4
Text pages 40–54
1.   Student presents Exercise
2.   Conditional Probability for Events that Occur in Stages; Probability Distributions for Independent Events (Sampling With Replacement)
3.   Experiment: Labyrinth
4.   Students present results
Lesson 4–5 Exercise: Labyrinth

Lesson 5
Text pages 54–66
1.   Student presents Exercise
2.   Distributions for Dependent Events (Sampling Without Replacement); Probability Histograms
3.   Experiment: Game Show
4.   Students present results
Lesson 5–6 Exercise: Calvin's Test

Lesson 6
Text pages 77–86
1.   Student presents Exercise
2.   Preview Experiment: Summarizing Ages
3.   Median; Five Number Summary; Stemplots
4.   Distribution of Students' Heights
5.   Experiment: Summarizing Ages
6.   Students present results
Lesson 6–7 Exercise: Summarizing and Comparing Distributions

Lesson 7
Text pages 87–95
1.   Student presents Exercise
2.   Mode & Mean; Statistics vs. Parameters
3.   Measure sample mean value of cards and sample proportion of diamonds
4.   Experiment: Verifying that Mean Equals Balance Point
5.   Students present results
Lesson 7–8 Exercise: Calculating Means

Lesson 8
Text pages 95–107
1.   Student presents Exercise

2.   Sample Mean and Sample Proportion as Estimates
3.   Experiment: Globe Toss
4.   Preview Experiment: Summarizing Grades
5.   Relative Frequency Histograms; Normal Distribution
6.   Experiment: Summarizing Report Card Grades
7.   Students present results
Lesson 8–9 Exercise: Histogram & Normal Curve for Boys'/Girls' Heights

Lesson 9

Text pages 111–122
1.   Student presents Exercise
2.   Sample Mean and Sample Proportion as Estimates; Standard Deviation; Distribution of Sample Mean and Proportion
3.   Experiment: Distribution of Sample Proportion of Red M&Ms™
4.   Experiment: Distribution of Sample Mean of Dice Rolls
5.   Students present results
Lesson 9–10 Exercise: Calculating Standard Deviations of Grades

Lesson 10

1.   Student presents Exercise
2.   Summarization; Awards and Certificates
3.   Party

An exercise will be assigned at the end of every class. Even if you can't solve it completely, I want you to do as much as you can and hand it in. An informal evaluation of students' performances made at the end of the two-week session will take their exercise work into account.

In addition, I encourage each of you to use your everyday experience, your creative imagination, and any other resources to make up and hand in problems (with solutions) related to the topics we cover. All of the good problems will be used in a problem-solving competition on the last day of class, with prizes given to the best table. The more good problems your table hands in, the better you'll do in our contest!

Special Materials Needed For Experiments

Counting Colors

stickers
one box each of six different colors, e.g. from office supply store,
or pens/crayons/markers of six different colors
double-sized paper or pairs of sheets taped together

Probabilities vs. Actual Proportions
cards
dice
dominoes
roulette wheel
containers
blocks: six blue and four not blue

Venn Diagrams
completed surveys

Sampling With or Without Replacement
four slips of paper
a container

Game Show
For each pair of students:
three colored envelopes numbered 1, 2, and 3
a piece of play money

"Doubtful" Events
a coin

Summarizing Ages
list of ages (in months) of class members

Mean Equals Balance Point
light rulers
small heavy blocks (e.g. Duplo™ blocks stacked in units of three together)
or sticks of butter or margarine
a small block or die for the base

Random Numbers
small slips of paper

Globe Toss
inflatable globe ball
atlas or geography book

Changing Your "Standard Deviation"
swivel chair

Sample Mean Dice Totals
four dice for each group

Sample Proportion of Red M&Ms
several bags of mixed color (preferably small baking) M&Ms™
a teaspoon
a tablespoon
a bowl

Relationships in Card Selections
decks of cards

Statistical Inference for Head Size
tape measure or string and ruler

Additional Materials
Dice and cards for ongoing experimentation
"combination" lock for discussion of its name
Lego™ magazine for discussion of mode
newspaper sports section for stemplot data

Class Survey

1. Are you male or female?
2. Is your hair light or dark?
3. Are your eyes brown or blue?
4. Do you have any brothers or sisters?
5. Do you have your own room?
6. Does your family have any pets?
7. Did you participate in organized sports (outside of school) this year? If so, which sports?
8. Did you attend lessons for a musical instrument this year? If so, which instrument?
9. Do you ordinarily take a bus to school?
10. What is your favorite subject in school?
11. What is your favorite color?
12. How many inches tall are you?
13. What is your foot length (barefoot, to the nearest quarter inch)?
14. How many months old are you?
15. What T-shirt size do you prefer to wear (S, M, L, or XL)?

Glossary

back-to-back stemplot—a display of two data sets to be compared, which depicts their patterns with a vertical list of stems and horizontal lists of leaves on either side of the stems

bias—the systematic tendency to over- or underestimate a parameter

box-and-whiskers plot—a graphical display of the five number summary

categorical variable—one which can take on various qualities, as opposed to quantities

combination—a selection from a group of distinct objects where order is not important

complement of an event—the event that it does not occur

conditional probability of a second event, given a first event—the probability that the second event occurs, given that the first event has occurred

confidence interval statement—one which tells how confident we are that an unknown parameter falls in a given interval

continuous random variable—one which has an infinite number of possible values over a continuous range.

data—facts, often numerical, about persons or things of interest

dependent events—those for which the probability of the occurrence of one is affected by whether or not the other occurs

deviation—the difference between a particular value of a variable and the mean value of that variable

direction of a relationship—tells how the size of values of one variable tends to relate to the size of values of the other variable

discrete random variable—one whose possible values can be counted out the way whole numbers can be counted out

expected value of a random variable—the sum of all products of each possible value with its probability

factorial of a whole number—the product of that number with all the whole numbers less than it, down to one; as a special case, zero factorial equals one

first quartile of a data set—the value such that one fourth of the data values are at or below it

five number summary of a data set—the list of its minimum, first quartile, median, third quartile, and maximum

form of a relationship—tells its shape, e.g., linear or curved

hypothesis—a claim about a parameter that is to be investigated or tested

independent events—those for which the probability of the occurrence of one is not affected by whether or not the other occurs

linear relationship—one whose scatterplot points fall close to a straight-line pattern

marginal distribution—one which concentrates on one variable at a time (when observations for two variables are given in a two-way table)

mean of a data set—the sum of all its values divided by the number of values

median of a data set—the middle value if the data set has an odd number of values; the average of the two middle values if there are an even number of values

mode of a data set—the value that occurs most frequently

moment of inertia—measures the energy required to start or stop a spinning body

negative relationship—one for which small values of one variable tend to occur with large values of the other, and vice versa

normal distribution—common pattern of behavior for variables which are centered at their mean, symmetric, and bell-shaped

numerical variable—one which takes on number values

outlier—an extreme value in a data set

parameter—a feature of an entire population, such as the mean of a population

of numerical values, or the proportion of a population of categorical values falling into a particular category

permutation—an arrangement of objects where the order is important

population—the entire group being studied

population mean value of a variable—(a parameter) the arithmetic average of all values of that variable

population proportion—(a parameter) the proportion of all population members falling into the category of interest

positive relationship—one for which small values of one variable tend to occur with small values of the other, and large values of one also tend to occur with large values of the other

probability—the study of random or chance behavior

probability distribution—the pattern of behavior of a random variable which tells its possible values and the probabilities of their occurrence

probability histogram—a display which represents probabilities by areas of rectangles

probability of an event—the likelihood, or chance, of its occurrence

random sample—one whose selection is based on chance alone

random variable—one whose values are numerical outcomes of a random or chance process

relative frequency histogram—a display which represents frequency of occurrence by areas of rectangles

sample—the part of a population which is examined in order to learn something about the entire population

sample mean—the random variable for the mean of a randomly selected sample, whose values vary according to the laws of chance

sample mean—the statistic which measures the average value of some numerical variable in a sample

sample proportion—the random variable for the proportion of members of a

randomly selected sample falling into a particular category

sample proportion—the statistic which measures the proportion of sample members falling into a particular category

scatterplot—a display of the relationship between two numerical variables

skewed distribution—an unbalanced one, with either smaller values [skewed left] or larger values [skewed right] tending to be further from the center

standard deviation—a measure of spread of a variable, calculated as the square root of the average squared difference of all values from their mean

statistic—something measured from data, such as sample mean or sample proportion

statistical inference—the process of using statistics to draw conclusions about unknown parameters

statistics—the science which concerns itself with the interpretation of data

stemplot—a display of a data set which depicts its pattern with a vertical list of stems followed horizontally by leaves

strength of a relationship between two variables—indicates how closely they are related to one another

symmetric distribution—one which is well-balanced on either side of its center

third quartile of a data set—the value such that three fourths of the data values are at or below it

tree diagram—a representation of various possibilities which uses branch-like arrangements of connected lines

variable—something which can take on different number or category values

Index

Other math titles from Prufrock Press: